CAMBRIDGE LIBRARY COLLECTION

Books of enduring scholarly value

Spiritualism and Esoteric Knowledge

Magic, superstition, the occult sciences and esoteric knowledge appear regularly in the history of ideas alongside more established academic disciplines such as philosophy, natural history and theology. Particularly fascinating are periods of rapid scientific advances such as the Renaissance or the nineteenth century which also see a burgeoning of interest in the paranormal among the educated elite. This series provides primary texts and secondary sources for social historians and cultural anthropologists working in these areas, and all who wish for a wider understanding of the diverse intellectual and spiritual movements that formed a backdrop to the academic and political achievements of their day. It ranges from works on Babylonian and Jewish magic in the ancient world, through studies of sixteenth-century topics such as Cornelius Agrippa and the rapid spread of Rosicrucianism, to nineteenth-century publications by Sir Walter Scott and Sir Arthur Conan Doyle. Subjects include astrology, mesmerism, spiritualism, theosophy, clairvoyance, and ghost-seeing, as described both by their adherents and by sceptics.

The Night Side of Nature

The novelist and children's author Catherine Crowe (c.1800–76) published *The Night Side of Nature* in two volumes in 1848. This lively collection of ghostly sketches and anecdotes was a Victorian best-seller and Crowe's most popular work. Sixteen editions appeared in six years, and it was translated into several European languages. The stories are intertwined with Crowe's own interpretations and commentaries which attack the scepticism of enlightenment thought and orthodox religion. Crowe seeks instead to encourage and re-invigorate a sense of wonder and mystery in life by emphasising the supernatural. The stories in Volume 1 centre on dreams, psychic presentiments, traces, wraiths, doppelgängers, apparitions, and imaginings of the after-life. Crowe's vivid tales, written with great energy and imagination, are classic examples of nineteenth-century spiritualist writing and strongly influenced other authors as well as providing inspiration for later adherents of ghost-seeing and psychic culture.

Cambridge University Press has long been a pioneer in the reissuing of out-of-print titles from its own backlist, producing digital reprints of books that are still sought after by scholars and students but could not be reprinted economically using traditional technology. The Cambridge Library Collection extends this activity to a wider range of books which are still of importance to researchers and professionals, either for the source material they contain, or as landmarks in the history of their academic discipline.

Drawing from the world-renowned collections in the Cambridge University Library, and guided by the advice of experts in each subject area, Cambridge University Press is using state-of-the-art scanning machines in its own Printing House to capture the content of each book selected for inclusion. The files are processed to give a consistently clear, crisp image, and the books finished to the high quality standard for which the Press is recognised around the world. The latest print-on-demand technology ensures that the books will remain available indefinitely, and that orders for single or multiple copies can quickly be supplied.

The Cambridge Library Collection will bring back to life books of enduring scholarly value (including out-of-copyright works originally issued by other publishers) across a wide range of disciplines in the humanities and social sciences and in science and technology.

The Night Side of Nature

Or, Ghosts and Ghost Seers

Volume 1

Catherine Crowe

CAMBRIDGE UNIVERSITY PRESS

Cambridge, New York, Melbourne, Madrid, Cape Town, Singapore,
São Paolo, Delhi, Dubai, Tokyo, Mexico City

Published in the United States of America by Cambridge University Press, New York

www.cambridge.org
Information on this title: www.cambridge.org/9781108027496

This edition first published 1848
This digitally printed version 2011

ISBN 978-1-108-02749-6 Paperback

THE

NIGHT SIDE OF NATURE;

OR,

GHOSTS AND GHOST SEERS.

BY

CATHERINE CROWE.

AUTHORESS OF

"SUSAN HOPLEY," "LILLY DAWSON,"

"ARISTODEMUS," &c. &c.

" Thou com'st in such a questionable shape,
That I will speak to thee!"

HAMLET.

IN TWO VOLUMES.

VOL. I.

LONDON:
T. C. NEWBY, 72, MORTIMER ST., CAVENDISH SQ.

1848.

INDEX.

VOL. I.

PREFACE.

IN my late novel of "Lilly Dawson," I announced my intention of publishing a work to be called "The Night Side of Nature;" this is it.

The term "Night Side of Nature" I borrow from the Germans, who derive it from the astronomers, the latter denominating that side of a planet which is turned from the sun, its *night side*. We are in this condition for a certain number of hours out of every twenty-four; and as, during this interval, external objects loom upon us but strangely and imperfectly, the Germans draw a parallel betwixt these vague and misty perceptions, and the similar obscure and uncertain glimpses we get of that veiled department of nature, of which, whilst comprising, as it does, the solution of questions concerning us more nearly than any other, we are yet in a state of entire and wilful ignorance. For science, at least science in this country, has put it aside as beneath her notice, because new facts that do not fit into old theories are troublesome, and not to be countenanced.

We are encompassed on all sides by wonders, and we can scarcely set our foot upon the ground, without trampling upon some marvellous production that our whole life and all our faculties would not suffice to comprehend. Familiarity, however, renders us insensible to the ordinary works of nature we are apt to forget the miracles they comprise, and even, sometimes, mistaking words for conceptions, commit the error of thinking we understand their mystery. But there is one class of these wonders with which, from their comparatively rare occurrence, we do not become familiar; and these, according to the character of the mind to which they are presented, are frequently either denied as ridiculous and impossible, or received as evidences of supernatural interference—interruptions of those general laws by which God governs the universe, which latter mistake arises from our only seeing these facts without the links that connect them with the rest of nature, just as in the faint light of a starlit night we might distinguish the tall mountains that lift their crests high into the sky, though we could not discern the low chain of hills that united them with each other.

There are two or three books, by German authors, entitled "The Night Side," or, "The Night Dominion of Nature," which are on subjects, more or less analagous to mine. Heinrick Schubert's is the most celebrated amongst them; it is a sort of cosmogony of the world, written in a spirit of philosophical mysticism—too much so for English readers, in general.

In undertaking to write a book on these subjects myself, I wholly disclaim the pretension of *teaching* or of enforcing opinions. My object is to suggest enquiry and stimulate observation, in order that we may endeavour, if possible, to discover something regarding our psychical nature, as it exists here in the flesh; and as it is to exist hereafter, out of it.

If I could only induce a few capable persons, instead of laughing at these things, to look at them, my object would be attained, and I should think my time well spent.

THE

NIGHT SIDE OF NATURE.

CHAPTER I.

INTRODUCTION.

"Know ye not that ye are the Temple of God, and that the Spirit of God dwelleth in you?"

1. *Cor.*, 3 c., 16 v.

Most persons are aware that the Greeks and Romans entertained certain notions regarding the state of the soul, or the immortal part of man, after the death of the body, which have been generally held to be purely mythological. Many of them, doubtless, are so; and of these I am not about to treat; but amongst their conceptions, there are some which, as they coincide with the opinions of many of the most enlightened persons of the present age, it may be desirable to consider more closely. I allude here particularly to their belief in the tripartite kingdom of the dead. According to this system, there were the Elysian fields, a region in which a certain sort of happiness was enjoyed; and Tartarus, the place of punishment for the wicked; each of which were, comparatively, but thinly inhabited. But there was, also, a mid-region, peopled with innumerable hosts of wandering and mournful spirits, who, although undergoing no torments, are represented as incessantly bewailing their condition, pining for the life they once enjoyed in the body, longing after the things of the earth, and occupying themselves with the same pursuits and objects, as had formerly constituted their business or their pleasure. Old habits are still dear to them, and they cannot snap the link that binds them to the earth.

Now, although we cannot believe in the existence of Charon, the three-headed dog, or Alecto, the serpent-haired fury, it may be worth while to consider whether the persuasion of the ancients with regard to that which concerns us all so nearly, namely, the destiny that awaits us when we have shaken off this mortal coil, may not have some foundation in truth: whether it might not be a remnant of a tradition transmitted from the earliest inhabitants of the earth, wrested by observation from nature, if not communicated from a higher source: and, also, whether circumstances of constant recurrence in all ages and in all nations, frequently observed and recorded by persons utterly ignorant of classical lore, and unacquainted, indeed, with the dogmas of any creed but their own, do not, as well as various passages in the Scriptures, afford a striking confirmation of this theory of a future life; whilst it, on the other hand, offers a natural and convenient explanation of their mystery.

To minds which can admit nothing but what can be explained and demonstrated, an investigation of this sort must appear perfectly idle; for whilst, on the one hand, the most acute intellect or the most powerful logic can throw little light on the subject, it is, at the same time—though I have a confident hope that this will not always be the case—equally irreducible within the present bounds of science; meanwhile, experience, observation, and intuition, must be our principal, if not our only guides. Because, in the seventeenth century, credulity outran reason and discretion; the eighteenth century, by a natural re-action, threw itself into an opposite extreme. Whoever closely observes the signs of the times, will be aware that another change is approaching. The contemptuous scepticism of the last age is yielding to a more humble spirit of enquiry; and there is a large class of persons amongst the most enlightened of the present, who are beginning to believe, that much which they had been taught to reject as fable, has been, in reality, ill-understood truth. Somewhat of the mystery of our own being, and of the mysteries that compass us about, are beginning to loom upon us—as yet, it is true, but obscurely; and, in the endeavour to follow out the clue they offer, we have but a feeble light to guide us. We must grope our way through the dim path before us, ever in danger of being led into error, whilst we may confidently reckon on being pursued by the shafts of ridicule—that weapon so easy to wield, so potent to the weak, so weak to the wise—which has delayed the births of so many truths, but never stifled one. The pharisaical scepticism which denies without investigation, is quite as perilous, and much more contemptible than the blind credulity which accepts all that it is caught without enquiry; it is, indeed, but another form of ignorance assuming to be knowledge And by *investigation*, I do not mean the hasty, captious, angry notice of an unwelcome fact, that too frequently claims the right of pro nouncing on a question; but the slow, modest, pains-taking examination, that is content to wait upon nature, and humbly follow out her disclosures, however opposed to pre-conceived theories or mortifying to human pride. If scientific men could but comprehend how they discredit the science, they really profess, by their despotic arrogance, and exclusive scepticism, they would surely, for the sake of that very science they love, affect more liberality and candour. This reflection, however, naturally suggests another, namely, do they really love science, or is it not too frequently with them but the means to an end? Were the love of science genuine, I suspect it would produce very different fruits to that which we see borne by the tree of knowledge, as it flourishes at present; and this suspicion is exceedingly strengthened by the recollection, that amongst the numerous students and professors of science I have at different times encountered, the real worshippers and genuine lovers of it, for its own sake, have all been men of the most single, candid, unprejudiced, and enquiring minds, willing to listen to all new suggestions, and investigate all new facts; not bold and self-sufficient, but humble and reverent suitors, aware of their own ignorance and unworthiness, and that they are yet but in the primer of nature's works, they do not permit themselves to pronounce upon her disclosures, or set limits to her decrees. They are content to admit that things new and unsuspected may yet be true; that their own knowledge of facts being extremely circumscribed, the systems attempted to be established on such uncertain data, must needs be very imperfect, and frequently altogether erroneous; and that it is therefore their duty, as it ought to be their pleasure, to welcome as a stranger every gleam of light that appears in the horizon, let it loom from whatever quarter it may.

But, alas! Poor science has few such lovers! *Les beaux yeux de sa cassette*, I fear, are much more frequently the objects of attraction than her own fair face.

The belief in a God, and in the immortality of what we call the soul, is common to all nations; but our own intellect does not enable us to form any conception of either one or the other. All the information we have on these subjects is comprised in such hints as the Scriptures here and there give us; whatever other conclusions we draw, must be the result of observation and experience. Unless founded upon these, the opinion of the most learned theologian, or the most profound student of science that ever lived, is worth no more than that of any other person. They know nothing whatever about these mysteries; and all *à priori* reasoning on them is utterly valueless. The only way, therefore, of attaining any glimpses of the truth in an enquiry of this nature, where our intellect can serve us so little, is to enter on it with the conviction that, knowing nothing, we are not entitled to reject any evidence that may be offered to us, till it has been thoroughly sifted, and proved to be fallacious. That the facts presented to our notice appear to us absurd, and altogether inconsistent with the notions our intellects would have enabled us to form, should have no weight whatever in the investigation. Our intellects are no measure of God Almighty's designs; and, I must say, that I do think one of the most irreverent, dangerous, and sinful things man or woman can be guilty of, is to reject with scorn and laughter any intimation which, however strangely it may strike upon our minds, and however adverse it may be to our opinions, may possibly be showing us the way to one of God's truths. Not knowing all the conditions, and wanting so many links of the chain, it is impossible for us to pronounce on what is probable and consistent, and what is not; and, this being the case, I think the time is ripe for drawing attention to certain phenomena, which, under whatever aspect we may consider them, are, beyond doubt, exceedingly interesting and curious; whilst, if the view many persons are disposed to take of them be the correct one, they are much more than this. I wish, also, to make the English public acquainted with the ideas entertained on these subjeets by a large proportion of German minds of the highest order. It is a distinctive characteristic of the thinkers of that country, that, in the first place, they do think independently and courageously; and, in the seeond, that they never shrink from promulgating the opinions they have been led to form, however new, strange, heterodox, or even absurd, they may appear to others. They do not succumb, as people do in this country, to the fear of ridicule; nor are they in danger of the odium that here pursues those who deviate from established notions; and, the consequence is, that, though many fallacious theories and untenable propositions may be advanced, a great deal of new truth is struck out from the collision; and in the result, as must always be the case, what is true lives and is established, and what is false dies and is forgotten. But here, in Britain, our critics and colleges are in such haste to strangle and put down every new discovery that does not emanate from themselves, or which is not a fulfilling of the ideas of the day, but which, being somewhat opposed to them, promises to be troublesome from requiring new thought to render it intelligible that one might be induced to suppose them divested of all confidence in this inviolable law; whilst the more important, and the higher the results involved may be, the more angry they are with those who advocate them. They do not quarrel with a new metal or a new plant, and even a new comet or a new island, stands a fair chance of being well received; the introduction of a planet appears, from late events, to be more difficult; whilst phrenology and mesmerism testify, that any discovery tending to throw light on what most deeply concerns us, namely, our own being, must be prepared to encounter a storm of angry persecution. And one of the evils of this hasty and precipitate opposition is, that the passions and interests of the opposers become involved in the dispute; instead of investigators, they become partisans; having declared against it in the outset, it is important to their petty interests that the thing shall not be true; and they determine that it *shall* not, if they can help it. Hence, these hasty, angry investigations of new facts, and the triumph with which failures are recorded; and hence the wilful overlooking of the axiom, that a thousand negatives cannot overthrow the evidence of one affirmative experiment. I always distrust those who have declared themselves strongly in the beginning of a controversy. Opinions which however rashly avowed, may have been honest at first, may have been changed for many a long day before they are retracted. In the mean time, the march of truth is obstructed, and its triumph is delayed; timid minds are alarmed; those who dare not, or cannot, think for themselves, are subdued; there is much needless suffering incurred, and much good lost; but the truth goes quietly on its way, and reaches the goal at last.

With respect to the subjects I am here going to treat of, it is not simply the result of my own reflections and convictions that I am about to offer. On the contrary, I intend to fortify my position by the opinions of many other writers; the chief of whom will, for the reasons above given, namely, that it is they who have principally attended to the question, be Germans. I am fully aware that in this country a very considerable number of persons lean to some of these opinions, and I think I might venture to assert that I have the majority on my side, as far as regards ghosts—for it is beyond a doubt that many more are disposed to believe than to confess—and those who do confess, are not few. The deep interest with which any narration of spiritual appearances bearing the stamp, or apparent stamp, of authenticty is listened to in every society, is one proof that, though the fear of ridicule may suppress, it cannot extinguish that intuitive persuasion, of which almost every one is more or less conscious.

I avow that, in writing this book, I have a higher aim than merely to afford amusement. I wish to engage the earnest attention of my readers; because I am satisfied that the opinions I am about to advocate, seriously entertained, would produce very beneficial results. We are all educated in the belief of a future state, but how vague and ineffective this belief is with the majority of persons, we too well know; for although, as I have said above, the number of those who are what is called believers in ghosts, and similar phenomena, is very large; it is a belief that they allow to sit extremely lightly on their minds. Although they feel that the evidence from within and from without is too strong to be altogether set aside, they have never permitted themselves to weigh the significance of the facts. They are afraid of that bugbear, Superstition—a title of opprobrium which it is very convenient to attach to whatever we do not believe ourselves. They forget that nobody has a right to call any belief superstitious, till he can prove that it is unfounded. Now, no one that lives can assert that the re-appearance of the dead is impossible; all he has a right to say is, that he does not believe it; and the interrogation that should immediately follow this declaration is, "Have you devoted your life to sifting all the evidence that has been adduced on the other side, from the earliest periods of history and tradition?" and even though the answer were in the affirmative, and that the investigation had been conscientiously pursued, it would be still a bold enquirer that would think himself entitled to say, the question was no longer open. But the rashness and levity with which mankind make professions of believing and disbelieving, are, all things considered, phenomena much more extraordinary than the most extraordinary ghost-story that was ever related. The truth is, that not one person in a thousand, in the proper sense of the word, believes anything; they only fancy they believe, because they have never seriously considered the meaning of the word and all that it involves. That which the human mind cannot conceive of, is apt to slip from its grasp like water from the hand; and life out of the flesh falls under this category. The observation of any phenomena, therefore, which enabled us to master the idea, must necessarily be extremely beneficial; and it must be remembered, that one single thoroughly well-established instance of the re-appearance of a deceased person, would not only have this effect, but that it would afford a demonstrative proof of the deepest of all our intuitions, namely, that a future life awaits us.

Not to mention the modern Germans of eminence, who have devoted themselves to this investigation, there have been men remarkable for intellect in all countries, who have considered the subject worthy of enquiry. Amongst the rest, Plato, Pliny, and Lucien; and in our own country, that good old divine,

Dr. Henry Moore, Dr. Johnson, Addison, Isaac Taylor, and many others. It may be objected that the eternally quoted case of Nicolai, the bookseller at Berlin, and Dr. Ferriar's "Theory of Apparitions," had not then settled the question; but nobody doubts that Nicolai's was a case of disease; and he was well aware of it himself, as it appears to me, everybody so afflicted, is. I was acquainted with a poor woman, in Edinborough, who suffered from this malady, brought on, I believe, by drinking; but she was perfectly conscious of the nature of the illusions; and that temperance and a doctor were the proper exorcists to lay the spirits. With respect to Dr. Ferriar's book, a more shallow one was assuredly never allowed to settle any question; and his own theory cannot, without the most violent straining, and the assistance of what he calls *coincidences*, meet even half the cases he himself adduces. That such a disease, as he describes, exists, nobody doubts; but I maintain that there are hundreds of cases on record, for which the explanation does not suffice; and if they have been instances of spectral illusion, all that remains to be said, is, that a fundamental reconstruction of the theory or that subject is demanded.

La Place says, in his "Essay on Probabilities," that "any case, however apparently incredible, if it be a recurrent case, is as much entitled under the laws of induction, to a fair valuation, as if it had been more probable before hand." Now, no one will deny that the case in question possesses this claim to investigation. Determined sceptics may, indeed, deny that there exists any well-authenticated instance of an apparition; but that, at present, can only be a mere matter of opinion; since many persons as competent to judge as themselves, maintain the contrary; and in the mean time, I arraign their right to make this objection till they have qualified themselves to do so, by a long course of patient and honest enquiry; always remembering that every instance of error or imposition discovered and adduced, has no positive value whatever in the argument, but as regards that single instance; though it may enforce upon us the necessity of strong evidence and careful investigation. With respect to the evidence, past and present, I must be allowed here to remark on the extreme difficulty of producing it. Not to mention the acknowledged carelessness of observers and the alleged incapacity of persons to distinguish betwixt reality and illusion, there is an exceeding shyness in most people, who, either have seen, or fancied they have seen, an apparition, to speak of it at all, except to some intimate friend; so that one gets most of the stories second-hand; whilst even those who are less chary of their communications, are imperative against their name and authority being given to the public. Besides this, there is a great tendency in most people, after the impression is over, to think they may have been deceived; and where there is no communication or other circumstance rendering this conviction impossible, it is not difficult to acquire it, or at least so much of it as leaves the case valueless. The seer is glad to find this refuge from the unpleasant feelings engendered; whilst surrounding friends, sometimes from genuine scepticism, and sometimes from good-nature, almost invariably lean to this explanation of the mystery. In consequence of these difficulties and those attending the very nature of the phenomena, I freely admit that the facts I shall adduce, as they now stand, can have no scientific value; they cannot in short enter into the region of science at all, still less into that of philosophy. Whatever conclusions we may be led to form, cannot be founded on pure induction. We must confine ourselves wholly within the region of opinion; if we venture beyond which, we shall assuredly founder. In the beginning, all sciences have been but a collection of facts, afterwards to be examined, compared, and weighed by intelligent minds. To the vulgar, who do not see the universal law which governs the universe, everything out of the ordinary course of events, is a prodigy; but to the enlightened mind there are no progidies; for it perceives that both in the moral and the physical world, there is a chain of uninterrupted connexion; and that the most strange and even apparently contradictory or supernatural fact or event will be found, on due investigation, to be strictly dependant on its antecedents. It is possible, that there may be a link wanting, and that our investigations may, consequently, be fruitless; but the link is assuredly there, although our imperfect knowledge and limited vision cannot find it

And it is here the proper place to observe, that, in undertaking to treat of the phenomena in question, I do not propose to consider them as supernatural; on the contrary, I am persuaded that the time will come, when they will be reduced strictly within the bounds of science. It was the tendency of the last age to reject and *deny* every thing they did not understand; I hope it is the growing tendency of the present one, to *examine* what we do not understand. Equally disposed with our predecessors of the eighteenth century to reject the supernatural, and to believe the order of nature inviolable, we are disposed to extend the bounds of nature and science, till they comprise within their limits all the phenomena, ordinary and extraordinary, by which we are surrounded. Scarcely a month passes, that we do not hear of some new and important discovery in science; it is a domain in which nothing is stable; and every year overthrows some of the hasty and premature theories of the preceding ones; and this will continue to be the case as long as scientific men occupy themselves each with his own subject, without studying the great and primal truths—what the French call *Les verités meres*—which link the whole together. Meantime, there is a continual unsettling. Truth, if it do not emanate from an acknowledged authority, is generally rejected; and error, if it do, is as often accepted; whilst, whoever disputes the received theory, whatever it be—we mean especially that adopted by the professors of colleges—does it at his peril. But there is a day yet brooding in the bosom of time, when the sciences will be no longer isolated; when we shall no longer deny, but be able to account for phenomena apparently prodigious; or have the modesty, if we cannot explain them, to admit that the difficulty arises solely from our own incapacity. The system of centralization in statistics, seems to be of doubtful advantage; but a greater degree of centralization appears to be very much needed in the domain of science. Some improvement in this respect might do wonders, particularly if reinforced with a slight infusion of patience and humility into the minds of scientific men; together with the recollection that facts and phenomena which do not depend on our will, must be waited for—that we must be at their command, for they will not be at ours.

But to return once more to our own subject. If we do believe that a future life awaits us, there can be nothing more natural than the desire to obtain some information as to what manner of life that is to be for which any one of us may, before this time to-morrow, have exchanged his present mode of being. That there does not exist a greater interest with regard to this question in the mind of man, arises, partly, from the vague intangible kind of belief he entertains of the fact; partly, from his absorption in worldly affairs, and the hard and indigestible food upon which his clerical shepherds pasture him—for, under dogmatic theology, religion seems to have withered away to the mere husk of spiritualism—and partly, also from the apparent impossibility of pursuing the enquiry to any purpose. As I said before, observation and experience can alone guide us in such an enquiry; for though most people have a more or less intuitive sense of their own immortality, intuition is silent as to the mode of it; and the question I am anxious here to discuss with my readers, is, whether we have any facts to observe, or any experience from which, on this most interesting of all subjects, a conclusion may be drawn. Great as the difficulty is of producing evidence, it will, I think, be pretty generally admitted, that, although each individual case, as it stands alone, may be comparatively valueless, the amount of recurrent cases forms a body of evidence, that on any other subject would scarcely be rejected; and since, if the facts are accepted, they imperatively demand an explanation—for, assuredly, the present theory of spectral illusions cannot comprise them—our enquiry, let it terminate in whatever conclusion it may, cannot be useless or uninteresting. Various views of the phenomeua in question may be taken; and although I shall offer my own opinions and the theories and opinions of others, I insist upon none; I do not write to dogmatise, but to suggest reflection and enquiry. The books of Dr. Ferriar, Dr. Hibbert, and Dr. Thatcher, the American, are all written to support one exclusive theory; and they only give such cases as serve to sustain it. They maintain that the whole phenomena are referrible to nervous or sanguineous derangement, and are mere subjective illusions; and whatever instance cannot be covered by this theory, they reject as false, or treat as a case of extraordinary coincidence. In short, they arrange the facts to their theory, not their theory to the facts. Their books cannot, therefore, claim to be considered as anything more than essays on a special disease; they have no pretence whatever to the character of investigations. The question, consequently, remains as much an open one as before they treated it; whilst we have the advantage of their experience and information, with regard to the peculiar malady that forms the subject of their works.

On that subject it is not my intention to enter; it is a strictly medical one, and every information may be obtained respecting it in the above-named treatises, and others, emanating from the faculty.

The subjects I do intend to treat of are the various kinds of prophetic dreams, presentiments, second-sight, and apparitions; and, in short, all that class of phenomena, which appears to throw some light on our physical nature, and on the probable state of the soul after death. In this discussion I shall make free use of my German authorities, Doctors Kerner, Stilling, Werner, Eschenmayer, Ennemoser, Passavent, Schubert, Von Meyer, &c. &c.; and I here make a general acknowledgement to that effect, because it would embarrass my book too much to be constantly giving names and references; although when I quote their words literally, I shall make a point of doing so; and because, also, that as I have been both thinking and reading much on these subjects for a considerable time past, I am, in fact, no longer in a condition to appropriate either to them or to myself, each his own. This, however, is a matter of very little consequence, as I am not desirous of claiming any ideas as mine that can be found elsewhere. It is enough for me, if I succeed in making a tolerably clear exposition of the subject, and can induce other people to reflect upon it.

CHAPTER II.

THE DWELLER IN THE TEMPLE.

It is almost needless to observe, that the Scriptures repeatedly speak of man as a tripartite being, consisting of spirit, soul, and body; and that, according to St. Paul, we have two bodies—a natural body, and a spiritual body; the former being designed as our means of communication with the external world—an instrument to be used and controlled by our nobler parts. It is this view of it, carried to a fanaticism, which has led to the various and extraordinary mortifications recorded of ascetics. As is remarked by the Rev. Hare Townshend, in a late edition of his book on Mesmerism, in this fleshly body consists our organic life; in the body which we are to retain through eternity, consists our fundamental life. May not the first, he says, "be a temporary development of the last, just as leaves, flowers and fruits, are the temporary developments of a tree. And in the same manner that these pass and drop away, yet leave the principle of reproduction behind, so may our present organs be detached from us by death, and yet the ground of our existence be spared to us continuously."

Without entering into the subtle disputes of philosophers, with regard to the spirit, a subject on which there is a standing controversy betwixt the disciples of Hegel, and those of other teachers, I need only observe that the Scriptures seem to indicate what some of the heathen sages taught, that the spirit that dwells within us is the spirit of God, incorporated in us for a period, for certain ends of his own, to be thereby wrought out. What those ends are, it does not belong to my present subject to consider. In this spirit so imparted to us, dwells, says Eschenmayer, the conscience, which keeps watch over the body and the soul, saying, "Thus shalt thou do!"

And it is to this Christ addresses himself when he bids his disciples become perfect, like their Father in Heaven. The soul is subject to the spirit; and its functions are, *to will*, or *choose*, *to think*, and *to feel*, and to become thereby cognizant of the true, the beautiful, and the good; comprehending the highest principle, the highest ideal, and the most perfect happiness. The *Ego*, or *I*, is the resultant of the three forces, Pneuma, Psyche, Soma—spirit, soul, and body.

In the sptrit or soul, or rather in both conjoined, dwells, also, the power of *spiritual seeing*, or *intuitive knowing;* for, as there is a spiritual body, there is a spiritual eye, and a spiritual ear, and so forth; or, to speak more correctly, all these sensuous functions are comprised in one universal sense, which does not need the aid of the bodily organs; but, on the contrary, is most efficient when most freed from them. It remains to be seen whether, or in what degree, such separation can take place during life; complete it cannot be till death; but whoever believes sincerely that the divine spirit dwells within him, can, I should think, find no difficulty in conceiving that, although from the temporary conditions towhich pt trahitis is subjected, this universal faculty is limited and obscured, it must still retain its indefeisible attribute.

We may naturally conclude that the most perfect state of man on earth consists in the most perfect unity of the spirit and the soul; and to those who in this life have attained the nearest to that unity, will the entire assimilation of the two, after they are separated from the body, be the easiest; whilst to those who have lived only their intellectual and external life, this union must be extremely difficult, the soul having chosen its part with the body and divorced itself, as much as in it lay, from the spirit. The voice of conscience is then scarcely heard; and the soul, degraded and debased, can no longer perform its functions of discerning the true, the beautiful, and the good.

On these distinct functions of the soul and spirit, however, it is not my intention to insist; since, it appears to me, a subject on which we are not yet in a condition to dogmatise. We know rather more about our bodies, by means of which the soul and spirit are united and brought into contact with the material world, and which are constructed wholly with a view to the conditions of that world; such as time, space, solidity, extension,

&c. &c. But we must conceive of God as necessarily independent of these conditions. To Him, all times and all places must be for ever present; and it is *thus* that he is omniscient and omnipresent; and since we are placed by the spirit in immediate relation with God and the spiritual world, just as we are placed by the body in immediate relation with the material world, we may, in the first place, form a notion of the possibility that some faint gleams of these inherent attributes may, at times, shoot up through the clay in which the spirit has taken up its temporary abode; and we may also admit, that through the connexion which exists betwixt us and the spiritual world, it is not impossible but that we may, at times, and under certain conditions, become cognizant of, and enter into more immediate relation with it. This is the only postulate I ask; for, as I said before, I do not wish to enforce opinions, but to suggest probabilities, or at least possibilities, and thus arouse reflection and enquiry.

With respect to the term *invisible world*, I beg to remind my readers, that what we call *seeing*, is merely the function of an organ constructed for that purpose, in relation to the external world; and so limited are its powers, that we are surrounded by many things in that world which we cannot see without the aid of artificial appliances, and many other things which we cannot see even with them; the atmosphere in which we live, for example, which, although its weight and mechanical forces are the subjects of accurate calculation, is entirely imperceptible to our visual organs. Thus, the fact that we do not commonly see them, forms no legitimate objection to the hypothesis of our being surrounded by a world of spirits, or of that world being inter-diffused amongst us. Supposing the question to be decided, that we do sometimes become cognizant of them, which, however, I admit it is not; since, whether the apparitions are subjective or objective, that is, whether they are the mere phenomena of disease, or real outstanding appearances, is the enquiry I desire to promote—but, I say, supposing that question were decided in the affirmative, the next that arises is, how, or by what means do we see them; or, if they address us, hear them? If that universal sense which appears to me to be inseparable from the idea of spirit, be once admitted, I think there can be no difficulty in answering this question; and if it be objected that we are conscious of no such sense, I answer that, both in dreams and in certain abnormal states of the body, it is frequently manifested. In order to render this more clear, and, at the same time, to give an interesting instance of this sort of phenomenon, I will transcribe a passage from a letter of St. Augustine to his friend Evadius (Epistola 159. Antwerp edition.)

"I will relate to you a circumstance," he writes, "which will furnish you matter for reflection. Our brother Sennadius, well known to us all as an eminent physician, and whom we especially love, who is now at Carthage, after having distinguished himself at Rome,and with whose piety and active benevolence you are well acquainted, could yet, nevertheless, as he has lately narrated to us, by any means bring himself to believe in a life after death. Now, God, doubtless, not willing that his soul should perish, there appeared to him, one night in a dream, a radiant youth of noble aspect, who bade him follow him; and as Sennadius obeyed, they came to a city where, on the right side, he heard a chorus of the most heavenly voices. As he desired to know whence this divine harmony proceeded, the youth told him that what he heard were the songs of the blessed; whereupon he awoke, and thought no more of his dream than people usually do.

On another night, however, behold! the youth appears to him again and asks if he knows him; and Sennadius related to him all the particulars of his former dream, which he well remembered. 'Then,' said the youth, 'was it whilst sleeping or waking that you saw these things?' 'I was sleeping,' answered Sennadius. 'You are right,' returned the youth, 'it was in your sleep that you saw these things; and know, oh Sennadius, that what you see now is also in your sleep. But if this be so, tell me where then is your body?' 'In my bed-chamber,' answered Sennadius. 'But know you not,' continued the stranger, 'that your eyes, which form a part of your body, are closed and inactive?' 'I know it,' answered he. 'Then,' said the youth, 'with what eyes see you these things?' And Sennadius could not answer him; and as he hesitated the youth spoke again, and explained to him the motive of his questions. 'As the eyes of your body,' said he, 'which lies now on your bed and sleeps, are inactive and useless, and yet you have eyes wherewith you see me and these things I have shown unto you, so after death, when these bodily organs fail you, you will have a vital power, whereby you will live; and a sensitive faculty, whereby you will perceive.

Doubt, therefore, no longer that there is a life after death.' And thus," said this excellent man, "was I convinced, and all doubts removed."

I confess there appears to me a beauty and a logical truth in this dream, that I think might convince more than the dreamer.

It is by the hypothesis of this universal sense, latent within us; an hypothesis which, whoever believes that we are immortal spirits, incorporated for a season in a material body, can scarcely reject, that I seek to explain those perceptions which are not comprised within the functions of our bodily organs. It seems to me to be the key to all, or nearly all, of them, as far as our own part in the phenomena extends. But, supposing this admitted, there would then remain the difficulty of accounting for the partial and capricious glimpses we get of it; whilst that department of the mystery which regards apparitions, except such as are the pure result of disease, we must grope our way, with very little light to guide us, as to the conditions and motives which might possibly bring them into any immediate relation with us.

To any one who has been fortunate enough to witness one genuine case of clairvoyance,

I think the conception of this universal sense will not be difficult; however, the mode of its exercise may remain utterly incomprehensible. As I have said above—to the great spirit and fountain of life, all things, both in space and time, must be present. However impossible it is to our finite minds to conceive this, we must believe it. It may, in some slight degree, facilitate the conception to remember, that action, once begun, never ceases—an impulse given is transmitted on for ever; a sound breathed reverberates in eternity; and thus the past *is* always present, although for the purpose of fitting us for this mortal life, our ordinary senses are so constituted as to be unperceptive of these phenomena. With respect to what we call *the future*, it is more difficult still for us to conceive it as present; nor, as far ar I know, can we borrow from the sciences the same assistance as mechanical discoveries have just furnished me with in regard to the past. How a spirit sees that which has not yet, to our senses, taken place, seems, certainly, inexplicable. *Foreseeing* it is not inexplicable; we foresee many things by arguing on given premises, although, from our own finite views, we are always liable to be mistaken. Louis Lambert says, "Such events as are the product of humanity, and the result of its intelligence, have their own causes, in which they lie latent, just as our actions are accomplished in our thoughts previous to any outward demonstration of them; presentiments and prophecies consist in the intuitive perception of these causes." This explanation which is quite conformable with that of Cicero, may aid us in some degree, as regards a certain small class of phenomena; but there is something involved in the question much more subtle than this. Our dreams can give us the only idea of it; for there we do actually see and hear, not only that which never was, but that which never will be. Actions and events, words and sounds, persons and places, are as clearly and vividly present to us, as if they were actually what they seem; and I should think that most people must be somewhat puzzled to decide in regard to certain scenes and circumstances that live in their memory, whether the images are the result of their waking or sleeping experience. Although by no means a dreamer, and without the most remote approximation to any faculty of presentiment, I know this is the case with myself. I remember, also, a very curious effect being produced upon me, when

I was abroad, soms years ago, from eating the unwholesome bread to which we were reduced, in consequence of a scarcity. Some five or six times a day I was seized with a sort of vertigo, during which I seemed to pass through certain scenes, and was conscious of certain words, which appeared to me to have a strange connexion, either with some former period of my life, or else some previous state of existence; the words and the scenes were on each occasion precisely the same. I was always aware of that; and I always made the strongest efforts to grasp and retain them in my memory; but I could not. I only knew that the thing *had been;* the words and the scenes were gone. I seemed to pass momentarily into another sphere and back again. This was purely the result of disorder; but, like a dream, it shows how we may be perceptive of that which is not, and which never may be; rendering it, therefore, possible to conceive that a spirit may be equally perceptive of that which shall be. I am very far from meaning to imply that these examples remove the difficulty; they do not explain the thing; they only show somewhat the mode of it. But it must be remembered that when physiologists pretend to settle the whole question of apparitions by the theory of spectral illusions, they are exactly in the same predicament. They can supply examples of similar phenomena; but how a person, perfectly in his senses, should receive the spectral visits, not only of friends, but strangers, when he is thinking of no such matter; or, by what process, mental or optical, the figures are conjured up, remains as much a mystery as before a line was written on the subject.

All people and all ages have believed, more or less, in prophetic dreams, presentiments, and apparitions; and all histories have furnished examples of them. That the truths may be frequently distorted and mingled with fable, is no argument against those traditions; if it were, all history must be rejected on the same plea. Both the Old and New Testament furnish numerous examples of these phenomena; and although Christ and the Apostles reproved all the superstitions of the age, these persuasions are not included in their reprehensions.

Neither is the comparitive rarity of these phenomena any argument against their possibility. There are many strange things which occur still more rarely, but which we do not look upon as supernatural or miraculous. Of nature's ordinary laws, we yet know but little;

of their aberrations and perturbations still less. How should we, when the world is a miracle and life a dream, of which we know neither the beginning nor the end! We do not even know that we see anything as it is; or rather, we know that we do not. We see things but as our visual organs represent them to us; and were those organs differently constructed, the aspect of the world, would to us, be changed. How, then, can we pretend to decide upon what is and what is not?

Nothing could be more perplexing to any one who read them with attention, than the trials for witchcraft of the seventeenth century Many of the feats of the ancient thaumaturgists and wonder-workers of the temples, might have been nearly as much so; but these were got rid of by the easy expedient of pronouncing them fables and impostures; but, during the witch mania, so many persons proved their faith in their own miraculous powers by the sacrifice of their lives, that it was scarcely possible to doubt their having some foundation for their own persuasion, though what that foundation could be, till the late discoveries in animal magnetism, it was difficult to conceive; but here we have a new page opened to us, which concerns both the history of the world and the history of man, as an individual; and we begin to see, that that which the ignorant thought supernatural, and the wise impossible, has been both natural and true. Whilst the scientific men of Great Britain, and several of our journalists, have been denying and ridiculing the reports of these phenomena, the most eminent physicians of Germany have been quietly studying and investigating them; and giving to the world, in their works, the results of their experience. Amongst the rest, Dr. Joseph Ennemoser, of Berlin, has presented to us in his two books on "Magic," and on "The Connexion of Magnetism with Nature and Religion," the fruits of his thirty years' study of this subject; during the course of which he has had repeated opportunities of investigating all the phenomena, and of making himself perfectly familiar with even the most rare and perplexing. To any one who has studied these works, the mysteries of the temples and of the witch trials, are mysteries no longer; and he writes with the professed design, not to make science mystical, but to bring the mysterious within the bounds of science. The phenomena, as he justly says, are as old as the human race. Animal magnetism is no new development, no new discovery. Inseparable from life, although, like many other vital phenomena, so subtle in its influences, that only in abnormal cases it attracts attention, it has exhibited itself more or less in all ages, and in all countries. But its value as a medical agent is only now beginning to dawn on the civilized world, whilst its importance, in a higher point of view, is yet perceived but by few. Every human being who has ever withdrawn himself from the strife, and the turmoil, and the distraction, of the world without, in order to look within, must have found himself perplexed by a thousand questions with regard to his own being, which he would find no one able to solve. In the study of animal magnetism, he will first obtain some gleams of a light which will show him that he is indeed the child of God! and that, though a dweller on the earth, and fallen, some traces of his divine descent, and of his unbroken connexion with a higher order of being, still remain to comfort and encourage him. He will find that there exists in his species the germs of faculties that are never fully unfolded here on earth, and which have no reference to this state of being. They exist in all men; but in most cases are so faintly elicited as not to be observable; and when they do shoot up here and there, they are denied, disowned, misinterpreted, and maligned. It is true, that their development is often the symptom and effect of disease, which seems to change the relations of our material and immaterial parts. It is true, that some of the phenomena resulting from these faculties are simulated by disease, as in the case of spectral illusions; and it is true, that imposture and folly intrude their unhallowed footsteps into this domain of science, as into that of all others; but there is a deep and holy well of truth to be discovered in this neglected bye-path of nature, by those who seek it, from which they may draw the purest consolations for the present, the most ennobling hopes for the future, and the most valuable aid in penetrating through the letter, into the spirit of the Scriptures.

I confess it makes me sorrowful when I hear men laughing, scorning, and denying this their brithright; and I cannot but grieve to think how closely and heavily their clay must be wrapt about them, and how the external and sensuous life must have prevailed over the internal, when no gleam from within breaks through to show them that these thinge are

CHAPTER III.

WAKING AND SLEEPING; AND HOW THE DWELLER IN THE TEMPLE SOMETIMES LOOKS ABROAD.

To begin with the most simple—or rather, I should say, the most ordinary, class of phenomena—for we can scarcely call that simple, the mystery of which we have never been able to penetrate—I mean dreaming—everybody's experience will suffice to satisfy them, that their ordinary dreams take place in a state of imperfect sleep; and that this imperfect sleep may be caused by any bodily or mental derangement whatever; or even from an ill-made bed, or too much or too little covering; and it is not difficult to conceive that the strange, confused, and disjointed visions we are subject to on these occasions, may proceed from some parts of the brain being less at rest than the others; so that, assuming phrenology to be fact, one organ is not in a state to correct the impressions of another. Of such vain and insignificant visions, I need scarcely say it is not my intention to treat; but, at the same time, I must observe, that when we have admitted the above explanation, as far as it goes, we have not, even in regard to *them*, made much progress towards removing the difficulty. If dreaming resembled thinking the explanations might be quite satisfactory; but the truth is, that dreaming is not thinking, as we think in our waking state; but is more analogous to thinking in delirium or acute mania, or in that chronic condition which gives rise to sensuous illusions. In our ordinary normal state, conceiving of places or persons does not enable us to see them or hold communion with them; nor do we fancy that we do either. It is true that I have heard some painters say, that by closing their eyes and concentrating their thoughts on an object, they can bring it more or less vividly before them; and Blake professed actually to see his sitters when they where not present; but whatever interpretations we may put upou this curious faculty, his case was clearly abnormal, and connected with some personal peculiarity, either physical or psychical; and, after making the most of it, it must be admitted that it can enter into no sort of comparison with that we possess in sleep, when, in our most ordinary dreams, untrammelled by time or space, we visit the uttermost ends of the earth, fly in the air, swim in the sea, listen to beautiful music and eloquent orations, behold the most charming, as well as the most loathsome objects; and not only see, but converse with our friends, absent or present, dead or alive. Every one, I think, will grant that there is the widest possible difference betwixt conceiving of these things when awake, and dreaming them. When we dream, we do, we see, we say, we hear, &c. &c., that is, we believe at the time that we do so; and what more can be said of us when we are awake, than that we *believe* we are doing, seeing, saying, hearing, &c. It is by external circumstances, and the results of our actions, that we are able to decide whether we have actually done a thing or seen a place, or only dreamt that we have done so; and as I have said above, after some lapse of time, we are not always able to distinguish between the two. Whilst dreaming, we frequently ask ourselves whether we are awake or asleep; and nothing is more common than to hear people say, "Well, I think I did, or heard, so and so; but I am not sure whether it was so, or whether I dreamt it." Thus, therefore, the very lowest order of dreaming, the most disjointed and perplexed, is far removed from the most vivid presentations of our waking thoughts; and it is in this respect, I think, that the explanations of the phenomena hitherto offered by phrenologists, and the metaphysicians of this country, are inadequate and unsatisfactory; whilst, as regards the analogy betwixt the visions of sleep and delirium, whatever similarity there may be in the effects, we cannot suppose the cause to be identical: since, in delirium, the images and delusions are the result of excessive action of the brain, which we must conclude to be the very reverse of its condition in sleep. Pinel certainly has hazarded an opinion that sleep is occasioned by an efflux of blood to the head, and consequent compression of the brain—a theory which would have greater weight were sleep more strictly periodical than it is; but which, at present, it seems impossible to reconcile with many established facts.

Some of the German physiologists and psychologists have taken a deeper view of this question of dreaming from considering it in connexion with the phenomena of animal magnetism; and although their theories differ in some respects, they all unite in looking towards that department of nature for instruction. Whilst one section of these enquirers, the Exegetical Society of Stockholm included, calls in the aid of supernatural agency, another, amongst whom Dr. Joseph Ennemoser, of Berlin, appears to be one of the most eminent, maintains that the explanation of the mystery is to be chiefly sought in the great and universal law of polarity, which extends not only beyond the limits of this earth, but beyond the limits of this system, which must necessarily be in connexion with all others; so that there is thus an eternal and never-ceasing inter-action, of which, from the multiplicity and contrariety of the influences we are insensible, just as we are insensible of the pressure of the atmoshere, from its impinging on us equally on all sides.

Waking and sleeping are the day and night sides of organic life, during which alternations an animal is placed in different relations to the external world, and to these alternations all organisms are subject. The completeness and independence of each individual organism, is in exact ratio to the number and completeness of the organs it developes; and thus the locomotive animal has the advantage of the plant or the zoophyte, whilst, of the animal kingdom, man is the most complete and independent; and, although still a member of the universal whole, and therefore incapable of isolating himself, yet better able than any other organism to ward off external influences, and comprise his world within himself. But, according to Dr. Ennemoser, one of the consequences of this very completeness, is a weak and insignificant development of instinct; and thus the healthy, waking, conscious man, is, of all organisms, the least sensible to the impressions of this universal intercommunication and polarity; although, at the same time, partaking of the nature of the plant and the animal, he is subject, like the first, to all manner of atmospheric, telluric, and periodic influences; and frequently exhibits, like the second, peculiar instinctive appetites and desires, and, in some individual organizations, very marked antipathies and susceptibilities with regard to certain objects and influences, even when not placed in any evident relation with them.

According to this theory, sleep is a retrograde step—a retreating into a lower sphere; in which condition, the sensuous functions being in abeyance, the instincts somewhat resume their sway. "In sleep and in sickness," he says, "the higher animals and man fall in a physico-organical point of view, from their individual independence, or power of self-sustainment; and their polar relation, that is, their relation to the healthy and waking man, becomes changed from a positive to a negative one; all men, in regard to each other, as well as all nature, being the subjects of this polarity. It is to be remembered, that this theory of Dr. Ennemoser's was promulgated before the discoveries of Baron von Reichenback in magnetism were made public, and the susceptibility to magnetic influences in the animal organism, which the experiments of the latter go to establish, is certainly in its favour; but whilst it pretends to explain the condition of the sleepers, and may possibly be of some service in our investigations into the mystery of dreaming, it leaves us as much in the dark as ever, with respect to the cause of our falling into this negative state; an enquiry in which little progress seems to have been hitherto made.

With respect to dreaming, Dr. Ennemoser

rejects the physiological theory, which maintains, that in sleep, magnetic or otherwise, the activity of the brain is transferred to the ganglionic system, and that the former falls into a subordinate relation. "Dreaming," he says, "is the gradual awakening of activity in the organs of imagination, whereby the presentation of sensuous objects to the spirit, which had been discontinued in profound sleep, is resumed. Dreaming," he adds, "also arises from the secret activity of the spirit in the innermost sensuous organs of the brain, busying the fancy with subjective sensuous images, the objective conscious day-life giving place to the creative dominion of the poetical genius, to which night becomes day, and universal nature its theatre of action; and thus the supersensuous or transcendent nature of the spirit becomes more manifest in dreaming than in the waking state. But, in considering these phenomena, man must be viewed both in his psychical and physical relations, and as equally subject to spiritual as to natural operations and influences; since, during the continuance of life, neither soul nor body can act quite independently of the other; for, although it be the immortal spirit which perceives, it is through the instrumentality of the sensuous organs that it does so; for of absolute spirit without body, we can form no conception."

What is here meant seems to be, that the, brain becomes the world to the spirit, before the impressions from the external world, do actually come streaming through by means of the external sensuous organs. The inner spiritual light illumines, till the outward, physical light overpowers and extinguishes it. But in this state, the brain, which is the store-house of acquired knowledge, is not in a condition to apply its acquisitions effectively; whilst the intuitive knowledge of the spirit, if the sleep be imperfect, is clouded by its interference.

Other physiologists, however, believe, from the numerous and well attested cases of the transference of the senses, in disease, to the pit of the stomach, that the activity of the brain in sleep *is* transferred to the epigastric region. The instances of this phenomenon, as related by Dr. Petetin and others, having been frequently published, I need not here quote. But, as Dr. Passavant observes, it is well known that the functions of the nerves differ in some animals; and that one set can supply the place of another; as in those cases where there is a great susceptibility to light, though no eyes can be discovered.

These physiologists believe, that, even during the most profouud sleep, the spirit retains its activity, a proposition which, indeed, we cannot doubt; "it wakes, though the senses sleep, retreating into its infinite depths, like the sun at night; living on its spiritual life undisturbed, whilst the body sinks into a state of vegetative tranquillity. Nor does it follow that the soul is unconscious in sleep, because in waking we have frequently lost all memory of its consciousness; since, by the repose of the sensuous organs, the bridge betwixt waking and sleeping is removed, and the recollections of one state are not carried into the other."

It will occur here to every one, how often in the instant of waking we are not only conscious that we have been dreaming, but are also conscious of the subject of the dream, which we try in vain to grasp, but which eludes us, and is gone for ever the moment we have passed into a state of complete wakefulness.

Now, with respect to this so called dreaming in profound sleep, it is a thing no one can well doubt, who thoroughly believes that his body is a temple built for the dwelling of an immortal spirit; for we cannot conceive of spirit sleeping, or needing that restoration which we know to be the condition of earthly organisms. If, therefore, the spirit wakes, may we not suppose that the more it is disentangled from the obstructions of the body, the more clear will be its perceptions; and that, therefore, in the profound natural sleep of the sensuous organs we may be in a state of clear-seeing. All who have attended to the subject are aware, that the clear-seeing of magnetic patients depends on the depth of their sleep; whatever circumstance, internal or external, tends to interrupt this profound repose of the sensuous organs, inevitably obscures their perceptions.

Again, with respect to the not carrying with us the recollections of one state into the other, should not this lead us to suspect, that sleeping and waking are two different spheres of existence; partaking of the nature of that *double life,* of which the records of human physiology have presented us with various instances, wherein a patient finds himself utterly divested of all recollection of past events and acquired knowledge, and has to begin life and education anew, till another transition takes place, wherein he recovers what he had lost, whilst he at the same time loses all he had lately gained, which he only recovers, once more, by another transition, restoring to him his lately acquired knowledge, but again obliterating his original stock, thus alternately passing from one state to the other, and disclosing a double life; an educated man in one condition, a child learning his alphabet in the next.

Where the transition from one state to another is complete, memory is entirely lost; but there are cases in which the change, being either gradual or modified, the recollections of one life are carried more or less into the other. We know this to be the case with magnetic sleepers, as it is with ordinary dreamers; and most persons have met with instances of the dream of one night being continued in the next. Treviranus mentions the case of a student who regularly began to talk the moment he fell asleep, the subject of his discourse being a dream, which he always took up at the exact point at which he had left it the previous morning. Of this dream he had never the slightest recollection in his waking state. A daughter of Sir George Mackenzie's, who died at an early age, was endowed with a remarkable genius for music, and was an accomplished organist. This young lady dreamt, during an illness, that she was at a party, where she had heard a new piece of music, which made so great an impression on her by its novelty and beauty, that, on awaking, she besought her attendants to bring her some paper, that she might write it down before she had forgotten it, an indulgence which, apprehensive of excitement, her medical attendant unfortunately forbad; for, apart from the additional psychological interest that would have been attached to the fact, the effects of compliance, judging from what ensued, would probably have been soothing, rather than otherwise. About ten days afterwards, she had a second dream, wherein she again found herself at a party, where she descried on the desk of a pianoforte, in a corner of the room, an open book, in which, with astonished delight, she recognised the same piece of music, which she immediately proceeded to play, and then awoke. The piece was not of a short or fugitive character, but in the style of an overture. The question, of course, remains, as to whether she was composing the music in her sleep, or by an act of clairvoyance, was perceiving some that actually existed. Either is possible, for, although she might have been incapable of composing so elaborate a piece in her waking state, there are many instances on record of persons performing intellectual feats in dreams, to which they were unequal when awake. A very eminent person assured me, that he had once composed some lines in his sleep, I think it was a sonnet, which far exceeded any of his waking performances of that description.

Somewhat analagous to this sort of double life, is the case of the young girl mentioned by Dr. Abercrombie and others, whose employment was keeping cattle, and who slept for some time, much to her own annoyance, in the room adjoining one occupied by an itinerant musician. The man, who played exceedingly well, being an enthusiast in his art, frequently practised the greater part of the night, performing on his violin very complicated, and difficult compositions, whilst the girl, so far from discovering any pleasure in his performances, complained bitterly of being kept awake by the noise. Some time after this, she fell ill and was removed to the house of a charitable lady, who undertook the charge of her; and here, by and by, the family were amazed by frequently hearing the most exquisite music in the night, which they at length discovered to proceed from the girl. The sounds were those of a violin, and the tuning and other preliminary processes were accurately imitated. She went through long and elaborate pieces, and afterwards was heard imitating, in the same way, the sounds of a pianoforte that was in the house. She also talked very cleverly on the subjects of religion and politics, and discussed, with great judgment, the characters and conduct of persons, public and private. Awake, she knew nothing of these things; but was, on the contrary, stupid, heavy, and had no taste whatever for music. Phrenology would probably interpret this phenomenon by saying, that the lower elements of the cerebral spinal axis, as organs of sensation, &c. &c., being asleep, the cluster of the higher organs requisite for the above combinations, were not only awake, but rendered more active from the repose of the others: but to me it appears, that we here see the inherent faculties of the spirit manifesting themselves, whilst the body slept. The same faculties must have existed when it was in a waking state; but the impressions and manifestations were then dependant on the activity and perfection of the sensuous organs, which seem to have been of an inferior order; and, consequently, no rays of this in-dwelling genius could pierce the coarse integument in which it was lodged.

Similar unexpected faculties have been not unfrequently manifested by the dying; and we may conclude to a certain degree from the same cause; namely, that the incipient death of the body is leaving the spirit more unobstructed. Dr Steinbech mentions the case of a clergyman, who, being summoned to administer the last sacraments to a dying peasant, found him, to his surprise, praying aloud in Greek and Hebrew, a mystery which could be no otherwise explained, than by the circumstance of his having, when a child, frequently heard the then minister of the parish praying in those languages. He had, however, never understood the prayers, nor indeed paid any attention to them; still less had he been aware that they lived in his memory. It would give much additional interest to this story had Dr. Steinbech mentioned how far the man, now, whilst uttering the words, understood their meaning; whether he was aware of what he was saying, or was only repeating the words by rote.

With regard to the extraordinary faculty of memory manifested in these and similar cases, I shall have some obervations to make in a subsequent part of this book.

Parallel instances are those of idiots, who, either in a somnambulic state, or immediately previous to death, have spoken as if inspired. At St. Jean de Maurienne, in Savoy, there was a dumb Cretin, who, having fallen into a natural state of somnambulism, not only was found to speak with ease, but also to the purpose; a faculty which disappeared, however, whenever he awoke. Dumb persons have likewise been known to speak when at the point of death.

The possibility of suggesting dreams to some sleepers by whispering in the ear, is a well known fact; but this can, doubtless, only be practicable where the sensuous organs are partly awake. Then, as with magnetic patients in a state of incomplete sleep, we have only reverie and imagination in place of clear-seeing.

The next class of dreams are those which partake of the nature of second sight, or prophecy, and of these there are various kinds; some being plain and literal in their premonitions, others allegorical and obscure; whilst some also regard the most unimportant, and others the most grave events of our lives. A gentleman engaged in business in the south of Scotland, for example, dreams that on entering his office in the morning, he sees seated on a certain stool, a person formerly in his service as clerk, of whom he had neither heard or thought for some time. He enquires the motive of the visit, and is told that such and such circumstances having brought the stranger to that part of the country, he could not forbear visiting his old quarters, expressing, at the same time, a wish to spend a few days in his former occupation, &c. &c. The gentleman, being struck with the vividness of the illusion, relates his dream at breakfast, and, to his surprise, on going to his office, there sits the man, and the dialogue that ensues is precisely that of the dream. I have heard of numerous instances of this kind of dream, where no previous expectation nor excitement of mind could be found to account for them, and where the fulfilment was too exact and literal, in all particulars, to admit of their being explained away by the ready resource of "an extraordinary coincidence." There are, also, on record, both in this country and others, many perfectly well authenticated cases of people obtaining prizes in the lottery, through having dreamt of the fortunate numbers. As many numbers, however, may have been dreamt of that were not drawn prizes, we can derive no conclusion from this circumstance.

A very remarkable instance of this kind of dreaming occurred a few years since to Mr. A. F., an eminent Scotch advocate, whilst staying in the neighbourhood of Lock Fyne, who dreamt one night that he saw a number of people in the street following a man to the scaffold. He discovered the features of the criminal in the cart distinctly; and, for some reason or other, which he could not account for, felt an extraordinary interest in his fate; insomuch that he joined the throng, and accompanied him to the place that was to terminate his earthly career. This interest was the more unaccountable, that the man had an exceedingly unprepossessing countenance, but it was, nevertheless, so vivid, as to induce the dreamer to ascend the scaffold, and address him, with a view to enable him to escape the impending catastrophe. Suddenly, however, whilst he was talking to him, the whole scene dissolved away, and the sleeper awoke. Being a good deal struck with the life-like reality of the vision, and the impression made on his mind by the features of this man, he related the circumstance to his friends, at breakfast, adding that he should know him anywhere, if he saw him. A few jests being made on the subject, the thing was forgotten.

On the afternoon of the same day, the advocate was informed that two men wanted to speak to him, and, on going into the hall, he was struck with amazement at perceiving that one of them was the hero of his dream!

"We are accused of a murder," said they, "and we wish to consult you. Three of us went out last night, in a boat; an accident has happened; our comrade is drowned, and they want to make us accountable for him." The advocate then put some interrogations to them, and the result produced in his mind, by their answers, was a conviction of their guilt. Probably the recollection of his dream rendered the effects of this conviction more palpable; for, one addressing the other, said, in Gaelic, "We have come to the wrong man; he is against us."

"There is a higher power than I against you," returned the gentleman; "and the only advice I can give you is, if you are guilty, fly immediately." Upon this, they went away; and the next thing he heard was, that they were taken into custody, on suspicion of the murder.

The account of the affair was, that, as they said, the three had gone out together on the preceding evening, and that in the morning, the body of one of them had been found on the shore, with a cut across his forehead. The father and friend of the victim had waited on the banks of the lake till the boat came in, and then demanded their companion; of whom, however, they professed themselves unable to give any account. Upon this, the old man led them to his cottage for the purpose of showing them the body of his son. One entered, and, at the sight of it, burst into a passion of tears: the other refused to do so, saying his business called him immediately home, and went sulkily away. This last was the man seen in the dream.

After a fortnight's incarceration, the former of these was liberated; and he then declared to the advocate his intention of bringing an action of damages for false imprisonment. He was advised not to do it. "Leave well alone," said the lawyer; "and if you'll take my advice, make off while you can." The man, however, refused to fly: he declared that he really did not know what had occasioned the death of his comrade. The latter had been at one end of the boat, and he at the other; when he looked round, he was gone; but whether he had fallen overboard, and cut his head as he fell, or whether he had been struck and pushed into the water, he did not know. The advocate became finally satisfied of this man's innocence; but the authorities, thinking it absurd to try one and not the other, again laid hands on him; and it fell to Mr. A. F. to be the defender of both. The difficulty was, not to separate their cases in his pleading; for, however morally convinced of the different ground on which they stood, his duty, professionally, was to obtain the acquittal of both; in which he finally succeeded, as regarded the charge of murder. They were, therefore, sentenced to two years' imprisonment; and, so far as the dream is concerned, here ends the story. There remains, however, a curious sequel to it.

A few years afterwards, the same gentleman being in a boat on Loch Fyne, in company with Sir T. D. L., happened to be mentioning these curious circumstances, when one of the boatmen said, that he "knew well about those two men; and that a very strange thing had occurred in regard to one of them." This one, on enquiry, proved to be the subject of the dream; and the strange thing was this: On being liberated, he had quitted that part of the country, and, in process of time, had gone to Greenock, and thence embarked in a vessel for Cork. But the vessel seemed fated never to reach its destination; one misfortune happened after another, till at length the sailors said, "This won't do; there must be a murderer on board with us." As is usual, when such a persuasion exists, they drew lots three times, and each time it fell on that man. He was, consequently, put on shore, and the vessel went on its way without him. What had become of him afterwards, was not known.

A friend of mine, being in London, dreamt that she saw her little boy playing on the terrace of her house in Northumberland, that he fell and hurt his arm, and she saw him lying apparently dead. The dream recurred two or three times, on the same night, and she awoke her husband, saying, she "feared something must have happened to Henry." In due course of post, a letter arrived from the governess, saying, that she was sorry to have to commuicate that, whilst playing on the terrace that morning, Master Henry had fallen over a heap of stones, and broken his arm; adding, that he had fainted after the accident, and had lain for some time insensible. The lady to whom this dream occurred, is not aware having ever manifested this faculty before or since.

Mrs. W. dreamt that she saw people ascending by a ladder to the chamber of her step-son, John; wakes, and says she is afraid he is dead, and that there was something odd in her dream about a watch and a candle. In the morning, a messenger is sent to enquire for the gentleman, and they find people ascending to his chamber window by a ladder, the door of the room being locked. They discover him dead on the floor, with his watch in his hand, and the candle between his feet. The same lady dreamt that she saw a friend in great agony; and that she heard him say, they were tearing his flesh from his bones. He was some time afterwards seized with inflammation, lay as she had seen him, and made use of those exact words.

A friend of mine dreamt, lately, that somebody said, her nephew must not be bled, as it would be dangerous. The young man was quite well, and there had been no design of bleeding him; but, on the following morning he had a tooth drawn, and an effusion of blood ensued, which lasted some days, and caused a good deal of uneasiness.

A farmer, in Worcestershire, dreamt that

his little boy, of twelve years old, had fallen from the waggon and was killed. The dream recurred three times in one night; but, unwilling to yield to superstitious fears, he allowed the child to accompany the waggoner to Kidderminster Fair. The driver was very fond of the boy, and he felt assured would take care of him; but, having occasion to go a little out of the road to leave a parcel, the man bade the child walk on with the waggon, and he would meet him at a certain spot. On arriving there, the horses were coming quietly forward, but the boy was not with them; and on retracing the road, he was found dead; having, apparently, fallen from the shafts and been crushed by the wheels.

A gentleman, who resided near one of the Scottish lakes, dreamt that he saw a number of persons surrounding a body, which had just been drawn out of the water. On approaching the spot, he perceives that it is himself, and the assistants are his own friends and retainers. Alarmed at the life-like reality of the vision, he resolved to elude the threatened destiny by never venturing on the lake again. On one occasion, however, it became quite indispensable that he should do so; and, as the day was quite calm, he yielded to the necessity, on condition that he should be put ashore at once on the opposite side, whilst the rest of the party proceeded to their destinations, where he would meet them. This was accordingly done: the boat skimmed gaily over the smooth waters, and arrived safely at the rendezvous, the gentlemen laughing at the superstition of their companion; whilst he stood smiling on the bank to receive them. But, alas! the fates were inexorable: the little promontory that supported him had been undermined by the water: it gave way beneath his feet, and life was extinct before he could be rescued. This circumstance was related to me by a friend of the family.

Mr. S. was the son of an Irish bishop, who set somewhat more value on the things of this world than became his function. He had always told his son that there was but one thing he could not forgive, and that was a bad marriage: meaning, by a bad marriage, a poor one. As cautions of this sort do not, by any means, prevent young people falling in love, Mr. S. fixed his affections on Lady O., a fair young widow, without any fortune; and, aware that it would be useless to apply for his father's consent, he married her without asking it. They were, consequently, exceedingly poor;

and, indeed, nearly all they had to live on was a small sinecure of forty pounds per annum, which Dean Swift procured for him. Whilst in this situation, Mr. S. dreamt one night that he was in the cathedral in which he had formerly been accustomed to attend service; that he saw a stranger, habited as a bishop, occupying his father's throne; and that, on applying to the verger for an explanation, the man said, that the bishop was dead, and that he had expired just as he was adding a codicil to his will in his son's favour. The impression made by the dream was so strong, that Mr. S. felt that he should have no repose till he had obtained news from home; and as the most speedy way of doing so, was to go there himself, he started on horseback, much against the advice of his wife, who attached no importance whatever to the circumstance. He had scarcely accomplished half his journey, when he met a courrier, bearing the intelligence of his father's death; and when he reached home, he found that there was a codicil attached to the will of the greatest importance to his own future prospects; but the old gentleman had expired, with the pen in his hand, just as he was about to sign it.

In this unhappy position, reduced to hopeless indigence, the friends of the young man proposed that he should present himself at the Vice-regal Palace, on the next levee day, in hopes that some interest might be excited in his favour; to which, with reluctance, he consented. As he was ascending the stairs, he was met by a gentleman whose dress indicated that he belonged to the Church.

"Good Heavens!" said he, to the friend who accompanied him, "Who is that?"

"That is Mr. ——, of so and so."

"Then he will be Bishop of L——!" returned Mr. S.; "For that is the man I saw occupying my father's throne!"

"Impossible!" replied the other; "he has no interest whatever, and has no more chance of being a bishop than I have."

"You will see," replied Mr. S., "I am certain he will.

They had made their obeisance above, and were returing, when there was a great cry without, and everybody rushed to the doors and windows to enquire what had happened. The horses attached to the carriage of a young nobleman had become restive, and were endangering the life of their master, when Mr. —— rushed forward, and, at the peril of his own, seized their heads, and afforded Lord

C. time to descend before they broke through all restraint, and dashed away. Through the interest of this nobleman and his friends, to whom Mr. —— had been previously quite unknown, he obtained the see of L. These circumstances were related to me by a member of the family.

It would be tedious to relate all the instances of this sort of dreaming which have come to my knowledge, but were they even much more rare than they are, and were there none of a graver and more mysterious kind, it might certainly occasion some surprise that they should have excited so little attention. When stories of this sort are narrated, they are listened to with wonder for the moment, and then forgotten, and few people reflect on the deep significance of the facts, nor the important consequences to us involved in the question, of how, with our limited faculties, which cannot foretel the events of the next moment, we should suddenly become prophets and seers.

The following dream, as it regards the fate of a very interesting person, and is, I believe, very little known, I will relate, though the story is of somewhat an old date:—Major Andre, the circumstances of whose lamented death are too well known to make it necessary for me to detail them here, was a friend of Miss Seward's, and, previously to his embarkation for America, he made a journey into Derbyshire, to pay her a visit, and it was arranged that they should ride over to see the wonders of the Peak, and introduce André to Newton, her minstrel, as she called him, and to Mr. Cunningham, the curate, who was also a poet.

Whilst these two gentlemen were awaiting the arrival of their guests, of whose intentions they had been apprised, Mr. Cunningham mentioned to Newton that, on the preceding night, he had had a very extraordinary dream, which he could not get out of his head. He had fancied himself in a forest; the place was strange to him; and, whilst looking about, he perceived a horseman approaching at great speed, who had scarcely reached the spot where the dreamer stood, when three men rushed out of the thicket, and, seizing his bridle, hurried him away, after closely searching his person. The countenance of the stranger being very interesting, the sympathy felt by the sleeper for his apparent misfortune awoke him; but he presently fell asleep again, and dreamt that he was standing near a great city, amongst thousands of people, and that he saw the same person he had seen seized in the wood brought out and suspended to a gallows. When André and Miss Seward arrived, he was horror-struck to perceive that his new acquaintance was the antitype of the man in the dream.

Mr. C., a friend of mine, told me, the other day, that he had dreamt he had gone to see a lady of his acquaintance, and that she had presented him with a purse. In the morning he mentioned the circumstance to his wife, adding that he wondered what should have made him dream of a person he had not been in any way led to think of; and, above all, that she should give him a purse. On that same day, a letter arrived from that lady to Mrs. C., containing a purse, of which she begged her acceptance. Here was the imperfect foreshadowing of the fact, probably from unsound sleep.

Another friend lately dreamt, one Thursday night, that he saw an acquaintance of his thrown from his horse; and that he was lying on the ground with the blood streaming from his face, which was much cut. He mentioned his dream in the morning, and being an entire disbeliever in such phenomena, he could not account for the impression made on his mind. This was so strong, that, on Saturday, he could not forbear calling at his friend's house; who, he was told, was in bed, having been thrown from his horse on the previous day, and much injured about the face.

Relations of this description having been more or less familiar to the world in all times and places; and the recurrence of the phenomena too frequent to admit of their reality being disputed, various theories were promulgated to account for them; and, indeed, there scarcely seems to be a philosopher or historian amongst the Greeks and Romans who does not make some allusion to this ill-understood department of nature; whilst, amongst the eastern nations, the faith in such mysterious revelations remains even yet undiminished. Spirits, good and evil, or the divinities of the heathen mythology, were generally called in to remove the difficulty; though some philosophers, rejecting this supernatural interference, sought the explanation in merely physical causes.

In the Druidical rites of the northern nations, women bore a considerable part: there were pristesses, who gave forth oracles and prophecies, much after the manner of the Pythonesses of the Grecian temples, and, no doubt, drawing their inspiration from the same sources; namely, from the influences of magnetism, and from narcotics. When the pure rites of Christianity superseded the Heathen forms of worship, tradition kept alive the memory of these vaticinations, together with some of the arcana of the Druidical groves; and hence, in the middle ages, arose a race of so-called witches and sorcerers, who were partly imposters, and partly self-deluded. Nobody thought of seeking the explanation of the facts they witnessed in natural causes; what had formerly been attributed to the influence of the Gods, was now attributed to the influence of the Devil; and a league with Satan was the universal solvent of all difficulties.

Persecution followed, of course; and men, women, and children, were offered up to the demon of superstition, till the candid and rational part of mankind, taking fright at the holocaust, began to put in their protest, and lead out a reaction, which, like all reactions, ran right into the opposite extreme. From believing everything, they ceased to believe anything; and, after swallowing unhesitatingly the most monstrous absurdities, they relieved themselves of the whole difficulty, by denying the plainest facts; whilst, what it was found impossible to deny, was referred to *imagination*—that most abused word, which explained nothing, but left the matter as obscure as it was before. Man's spiritual nature was forgotten; and what the senses could not apprehend, nor the understanding account for, was pronounced to be impossible. Thank God! we have lived through that age, and, in spite of the struggles of the materialistic school, we are fast advancing to a better. The traditions of the saints who suffered the most appalling tortures, and slept or smiled the while, can scarcely be rejected now, when we are daily hearing of people undergoing frightful operations, either in a state of insensibility, or whilst they believe themselves revelling in delight; nor can the psychological intimations which these facts offer, be much longer overlooked. One revelation must lead to another; and the wise men of the world will, ere long, be obliged to give in their adherence to Shakspere's much quoted axiom, and confess that "there *are* more things in Heaven and earth than are dreamt of in their philosophy!"

CHAPTER IV.

ALLEGORICAL DREAMS, PRESENTIMENT, ETC.

It has been the opinion of many philosophers, both ancient and modern, that, in the original state of man, as he came forth from the hands of his Creator, that knowledge which is now acquired by pains and labour, was intuitive. His material body was given him for the purpose of placing him in relation with the material world, and his sensuous organs for the perception of material objects, but his soul was a mirror of the universe, in which everything was reflected, and, probably, is so still, but that the spirit is no longer in a condition to perceive it. Degraded in his nature, and distracted by the multiplicity of the objects and interests that surround him, man has lost his faculty of spiritual seeing; but in sleep, when the body is in a state of passivity, and external objects are excluded from us by the shutting up of the senses through which we perceive them, the spirit, to a certain degree, freed from its impediments, may enjoy somewhat of its original privilege. "The soul, which is designed as the mirror of a superior spiritual order" (to which it belongs), still receives, in dreams, some rays from above, and enjoys a foretaste of its future condition; and, whatever interpretation may be put upon the history of the Fall, few will doubt that, before it, man must have stood in a much more intimate relation to his Creator than he has done since. If we admit this, and that, for the above hinted reasons, the soul in sleep may be able to exercise somewhat of its original endowment, the possibility of what is called prophetic dreaming may be better understood,

"Seeing in dreams," says Ennemoser, "is a self-illumining of things, places, and times;" for relations of time and space form no obstruction to the dreamer: things, near and far, are alike seen in the mirror of the soul, according to the connexion in which they stand to each other; and, as the future is but an unfolding of the present, as the present is of the past, one being necessarily involved in the other, it is not more difficult to the untrammelled spirit to perceive what *is* to happen, than what has *already* happened. Under what peculiar circumstances it is that the body and soul fall into this particular relative condition, we do not know; but that certain families and constitutions are more prone to these conditions than others, all experience goes to establish. According to the theory of Dr. Ennemoser, we should conclude that they are more susceptible to magnetic influences, and that the body falls into a more complete state of negative polarity.

In the histories of the Old Testament, we constantly find instances of prophetic dreaming, and the voice of God was chiefly heard by the prophets in sleep; seeming to establish that man is, in that state, more susceptible of spiritual communion, although the being thus made the special organ of the divine will, is altogether a different thing from the mere disfranchisement of the embodied spirit in ordinary cases of clear seeing in sleep. Profane history, also, furnishes us with various instances of prophetic dreaming, which it is unnecessary for me to refer to here. But there is one thing very worthy of remark, namely, that the allegorical character of many of the dreams recorded in the Old Testament, occasionally pervades those of the present day. I have heard of several of this nature, and Oberlin, the good pastor of Ban de la Roche, was so subject to them, that he fancied he had acquired the art of interpreting the symbols. This characteristic of dreaming is in strict conformity with the language of the Old Testament, and of the most ancient nations. Poets and prophets, heathen and Christian, alike express themselves symbolically, and, if we believe that this language prevailed in the early ages of the world, before the external and intellectual life had predominated over the instinctive and emotional, we must conclude it to be the natural language of man, who must, therefore, have been gifted with a conformable faculty, of comprehending these hieroglyphics; and hence it arose that the interpreting of dreams became a legitimate art. Long after these instinctive faculties were lost, or rather obscured, by the turmoil and distractions of sensuous life, the memories and traditions of them remained, and hence the superstructure of jugglery and imposture that ensued, of which the gipsies form a signal example, in whom, however, there can be no doubt, that some occasional gleams of this original endowment may still be found, as is the case, though more rarely, in individuals of all races and conditions. The whole of nature is one large book of symbols, which, because we have lost the key to it, we cannot decipher. "To the first man," says Hamann, "whatever his ear heard, his eye saw, or his hand touched, was a living word; with this word in his heart and in his mouth, the formation of language was easy. Man saw things in their essence and properties, and named them accordingly.

There can be no doubt that the heathen forms of worship and systems of religion were but the external symbols of some deep meanings, and not the idle fables that they have been too frequently considered; and it is absurd to suppose that the theology which satisfied so many great minds, had no better foundation than a child's fairy tale.

A maid servant, who resided many years in a distinguished family in Edinburgh, was repeatedly warned of the approaching death of certain members of that family, by dreaming that one of the walls of the house had fallen

Shortly before the head of the family sickened and died, she said she had dreamt that the main wall had fallen.

A singular circumstance which occurred in this same family, from a member of whom I heard it, is mentioned by Dr. Abercrombie. On this occasion the dream was not only prophetic, but the symbol was actually translated into fact.

One of the sons being indisposed with a sore throat, a sister dreamt that a watch, of considerable value, which she had borrowed from a friend, had stopt; that she had awakened another sister and mentioned the circumstance, who answered that "something much worse had happened, for Charles's breath had stopt." She then awoke, in extreme alarm, and mentioned the dream to her sister, who, to tranquillize her mind, arose and went to the brother's room, where she found him asleep and the watch going. The next night, the same dream recurred, and the brother was again found asleep and the watch going. On the following morning, however, this lady was writing a note in the drawing-room, with the watch beside her, when, on taking it up, she perceived it had stopt; and she was just on the point of calling her sister to mention the circumstance, when she heard a scream from her brother's room, and the sister rushed in with the tidings that he had just expired. The malady had not been thought serious; but a sudden fit of suffocation had unexpectedly proved fatal.

This case, which is established beyond all controversy, is extremely curious in many points of view: the acting out of the symbol, especially. Symbolical events of this description have been often related, and as often laughed at. It is easy to laugh at what we do not understand; and it gives us the advantage of making the timid narrator ashamed of his fact, so that if he do not wholly suppress it, he at least ensures himself by laughing, too, the next time he relates it. It is said that Goethe's clock stopt at the moment he died; and I have heard repeated instances of this strange kind of synchronism, or magnetism, if it be by magnetism that we are to account for the mystery. One was told me very lately by a gentleman to whom the circumstance occurred.

On the 16th of August, 1769, Frederick II. of Prussia is said to have dreamt, that a star fell from heaven, and occasioned such an extraordinary glare that he could with great difficulty find his way through it. He mentioned the dream to his attendants, and it was afterwards observed that it was on that day Napoleon was born.

A lady, not long since, related to me the following circumstance:—Her mother, who was at the time residing in Edinburgh, in a house, one side of which looked into a wynd, whilst the door was in the High-street, dreamt that, it being Sunday morning, she had heard a sound, which had attracted her to the window; and, whilst looking out, had dropt a ring from her finger into the wynd below. That she had, thereupon, gone down in her night clothes to seek it; but when she reached the spot, it was not to be found. Returning, extremely vexed at her loss, as she re-entered her own door she met a respectable looking young man, carrying some loaves of bread. On expressing her astonishment at finding a stranger there at so unseasonable an hour, he answered, by expressing his at seeing her in such a situation. She said she had dropt her ring, and had been round the corner to seek it; whereupon, to her delighted surprise, he presented her with her lost treasure. Some months afterwards, being at a party, she recognised the young man seen in her dream, and learnt that he was a baker. He took no particular notice of her on that occasion; and, I think, two years elapsed be fore she met him again. This second meeting, however, led to an acquaintance, which terminated in marriage.

Here the ring and the bread are curiously emblematic of the marriage, and the occupation of the future husband.

Miss L., residing at Dalkeith, dreamt that her brother, who was ill, called her to his bedside, and gave her a letter, which he desired her to carry to their aunt, Mrs. H., with the request that she would deliver it to John," (John was another brother, who had died previously, and Mrs. H. was at the time ill.) He added that, "he himself was going *there* also, but that Mrs. H. would be *there* before him." Accordingly, Miss L. went, in her dream, with the letter to Mrs. H., whom she found dressed in white, and looking quite radiant and happy. She took the letter, saying she was going *there* directly, and would deliver it.

On the following morning, Miss L. learnt that her aunt had died in the night. The brother died some little time afterwards.

A gentleman who had been a short time visiting Edinburgh, was troubled with a cough, which, though it occasioned him no alarm, he resolved to go home to nurse. On the first night of his arrival, he dreamt that one half of the house was blown away. His bailiff, who resided at a distance, dreamt the same dream on the same night. The gentleman died within a few weeks.

"This symbolical language which the Deity appears to have used" (witness Peter's dream, Acts ii., and others,) "in all his revelations to man, is in the highest degree, what poetry is in a lower; and the language of dreams, in the lowest, namely, the original natural language of man; and we may fairly ask whether this language, which here plays an inferior part, be not, possibly, the proper language of a higher sphere, whilst we, who vainly think ourselves awake, are, in reality, buried in a deep, deep sleep, in which, like dreamers who imperfectly hear the voices of those around them, we occasionally apprehend, though obscurely, a few words of this Divine tongue." (*Vide Schubert.*)

This subject of sleeping and waking is a very curious one, and might give rise to strange questionings. In the case of those patients above mentioned, who seem to have two different spheres of existence, who shall say which is the waking one, or whether either of them be so? The speculations of

Mr. Dove on this subject merited more attention, I think, than they met with, when he lectured in Edinburgh. He maintained that, long before he had paid any attention to magnetism, he had arrived at the conclusion that there are as many states or conditions of mind beyond sleep, as there are on this side of it; passing through the different stages of dreaming, reverie, contemplation, &c., up to perfect vigilance. However this be, in this world of appearance, where we see nothing as it is, and where, both as regards our moral and physical relations, we live in a state of continual delusion, it is impossible for us to pronounce on this question. It is a common remark, that some people seem to live in a dream, and never to be quite awake; and the most cursory observer cannot fail to have been struck with examples of persons in this condition, especially in the aged.

With respect to this allegorical language, Ennemoser observes that, "since no dreamer learns it of another, and still less from those who are awake, it must be natural to all men." How different, too, is its comprehensiveness and rapidity, to our ordinary language! We are accustomed, and with justice, to wonder at the admirable mechanism by which, without fatigue or exertion, we communicate with our fellow beings; but how slow and ineffective is human speech, compared to this spiritual picture-language, where a whole history is understood at a glance! and scenes that seem to occupy days and weeks, are acted out in ten minutes. It is remarkable that this hieroglyphic language appears to be the same amongst all people; and that the dream interpreters of all countries construe the signs alike. Thus, the dreaming of deep water denotes trouble, and pearls are a sign of tears.

I have heard of a lady, who, whenever a misfortune was impending, dreamt that she saw a large fish. One night, she dreamt that this fish had bitten two of her little boy's fingers. Immediately afterwards, a school-fellow of the child's injured those two very fingers, by striking him with a hatchet; and I have met with several persons who have learnt, by experience, to consider one particular dream as the certain prognostic of misfortune.

A lady, who had left the West Indies when six years old, came one night, fourteen years afterwards, to her sister's bed-side, and said, "I know uncle is dead. I have dreamt that I saw a number of slaves in the large storeroom at Barbadoes, with long brooms sweeping down immense cobwebs. I complained to my aunt, and she covered her face and said, "Yes, he is no sooner gone than they disobey him." It was afterwards ascertained, that Mr. P. had died on that night; and that he had never permitted the cobwebs in this room to be swept away, of which, however, the lady assures me she knew nothing; nor could she or her friends conceive what was meant by the symbol of the cobwebs, till they received the explanation subsequently, from a member of the family.

The following very curious allegorical dream I give, not in the words of the dreamer, but in those of her son, who bears a name destined, I trust, to a long immortality:—

"Wooer's Abbey-Cottage,
"Dunfermiline-in-the Woods,
"Monday Morning, 31st May, 1847.

"Dear Mrs. Crowe,

"*That* dream of my mother's was as follows:—She stood in a long, dark, empty gallery: on her one side was my father, and on the other my eldest sister Amelia; then myself, and the rest of the family, according to their ages. At the foot of the hall stood my youngest sister Alexes, and above her my sister Catherine—a creature, by the way, in person and mind more like an angel of heaven than an inhabitant of earth. We all stood silent and motionless. At last *It* entered—the unimagined *something* that, casting its grim shadow before, had enveloped all the trivialties of the preceding dream in the stifling atmosphere of terror. It entered, stealthily descending the three steps that led from the entrance down into the chamber of horror: and my mother *felt It was Death.* He was dwarfish, bent, and shrivelled. He carried on his shoulder a heavy axe; and had come, she thought, to destroy 'all her little ones at one fell swoop.' On the entrance of the shape my sister Alexes leapt out of the rank, interposing herself between him and my mother. He raised his axe and aimed a blow at Catherine: a blow which, to her horror, my mother could not intercept; though she had snatched up a three-legged stool, the sole furniture of the apartment, for that purpose. She could not, she felt, fling the stool at the figure without destroying Alexes, who kept shooting out and in between her and the ghastly thing. She tried, in vain, to scream; she besought my father, in agony, to avert the impending stroke; but he did not hear, or did not heed her; and stood motionless, as in a trance. Down came the axe, and poor Catherine fell in her blood, cloven to 'the white halse bane.' Again the axe was lifted, by the inexorable shadow, over the head of my brother, who stood next in the line. Alexes had somewhere disappeared behind the ghastly visitant; and, with a scream, my mother flung the footstool at his head. He vanished, and she awoke. This dream left on my mother's mind a fearful apprehension of impending misfortune, 'which would not pass away.' It was *murder* she feared; and her suspicions were not allayed by the discovery that a man—some time before discarded by my father for bad conduct, and with whom she had, somehow, associated the *Death* of her dream—had been lurking about the place, and sleeping in an adjoining outhouse on the night it occurred, and for some nights previous and subsequent to it. Her terror increased. Sleep forsook her; and every night, when the house was still, she arose and stole, sometimes with a candle, sometimes in the dark, from room to room, listening, in a sort of waking night-mare, for the breathing of the assassin, who, she imagined, was lurking in some one of them. This could not last. She reasoned with herself; but her terror became intolerable, and she related her dream to my father, who, of course, called her a fool for her pains—whatever might be his real opinion of the matter. Three months had elapsed, when we, children, were all of us seized with scarlet fever. My sister Catherine died almost immediately—sacrificed, as my mother, in her misery, thought, to her (my mother's) overanxiety for Alexes, whose danger seemed more imminent. The dream-prophecy was in part fulfilled. I, also, was at death's door—given up by the doctors, but not by my mother: she was confident of my recovery; but for my brother, who was scarcely considered in danger at all, but on whose head *she had seen* the visionary axe impending, her fears were great; for she could not recollect whether the blow had, or had not, descended when the spectre vanished. My brother recovered, but relapsed, and barely escaped with life; but Alexes did not. For a year and ten months the poor child lingered; and almost every night I had to sing her asleep; often, I remember, through bitter tears; for I knew she was dying, and I loved her the more as she wasted away. I held her little hand as she died; I followed her to the grave—the last thing that I have *loved* on earth. And *the dream was fulfilled.*

"True and sincerely your's,
"J. NOEL PATON."

The dreaming of coffins and funerals, when a death is impending, must be considered as examples of this allegorical language. Instances of this kind are extremely numerous. Not unfrequently the dreamer, as in cases of second sight, sees either the body in the coffin, so as to be conscious of who is to die; or else, is made aware of it from seeing the funeral procession at a certain house, or from some other significant circumstance. This faculty which has been supposed to belong peculiarly to the Highlanders of Scotland, appears to be fully as well known in Wales and on the continent, especially in Germany.

The language of dreams, however, is not always symbolical. Occasionally, the scene that is transacting at a distance, or that is to be transacted at some future period, is literally presented to the sleeper, as things appear to be presented in many cases of second sight, and also in clairvoyance; and, since we suppose him, that is, the sleeper, to be in a temporarily magnetic state, we must conclude that the degree of perspecuity, or translucency of the vision, depends on the degree of that state. Nevertheless, there are considerable difficulties attending this theory. A great proportion of the prophetic dreams we hear of, are connected with the death of some friend or relative. Some, it is true, regard unimportant matters as visits, and so forth; but this is generally, though not exclusively, the case only with persons who have a constitutional tendency to this kind of dreaming, and with whom it is frequent; but it is not uncommon for those who have not discovered any such tendency, to be made aware of a death; and the number of dreams of this description I meet with, is very considerable. Now, it is difficult to conceive what the condition is, that causes this perception of an approaching death; or why, supposing, as we have suggested above, that, when the senses sleep, the untrammelled spirit *sees*, the memory of this revelation, if I may so call it, so much more frequently survives than any other, unless, indeed, it be the force of the shock sustained, which shock, it is to be remarked, always wakes the sleeper; and this may be the reason that, if he fall asleep again, the dream is almost invariably repeated.

I could fill pages with dreams of this description which have come to my knowledge, or been recorded by others.

Mr. H., a gentleman with whom I am acquainted, a man engaged in active business, and apparently as little likely as any one I ever knew to be troubled with a faculty of this sort, dreamt that he saw a certain friend of his dead. The dream was so like reality, that, although he had no reason whatever to suppose his friend ill, he could not forbear sending in the morning to enquire for him. The answer returned was, that Mr. A. was out, and was quite well. The impression, however, was so vivid, that, although he had nearly three miles to send, Mr. H. felt that he could not start for Glasgow, whither business called him, without making another enquiry. This time his friend was at home, and answered for himself, that he was very well, and that somebody must have been hoaxing H., and making him believe otherwise. Mr. H. set out on his journey, wondering at his own anxiety, but unable to conquer it. He was absent but a few days—I think, three; and the first news he heard on his return was, that his friend had been seized with an attack of inflammation, and was dead.

A German professor lately related to a friend of mine, that, being some distance from home, he dreamt that his father was dying, and was calling for him. The dream being repeated, he was so far impressed as to alter his plans, and return home, where he arrived in time to receive his parent's last breath. He was informed that the dying man had been calling upon his name repeatedly, in deep anguish at his absence.

A parallel case to this is that of Mr. R. E. S., an accountant in Edinburgh, and a shrewd man of business, who relates the following circumstance as occurring to himself. He is a native of Dalkeith, and was residing there, when, being about fifteen years of age, he left home on a Saturday, to spend a few days with a friend at Prestonpans. On the Sunday night, he dreamt that his mother was extremely ill, and started out of his sleep with an impression that he must go to her immediately. He even got out of bed with the intention of doing so, but, reflecting that he had left her quite well, and that it was only a dream, he returned to bed, and again fell asleep. But the dream returned, and, unable longer to control his anxiety, he arose, dressed himself in the dark, quitted the house, leaping the railings that surrounded it, and made the best of his way to Dalkeith. On reaching home, which he did before daylight, he tapped at the kitchen window, and, on gaining admittance, was informed that on the Saturday evening, after he had departed, his mother had been seized with an attack of British cholera, and was lying above, extremely ill. She had been lamenting his absence extremely, and had scarcely ceased crying, "Oh, Ralph, Ralph! what a grief that you are away!"

At nine o'clock he was admitted to her room; but she was no longer in a condition to recognise him, and she died within a day or two. Instances of this sort are numerous; but it would be tedious to narrate them, especially as there is little room for variety in the details. I shall, therefore, content myself with giving one or two specimens of each class, confining my examples to such as have been communicated to myself, except where any case of particular interest leads me to deviate from this plan. The frequency of such phenomena may be imagined, when I mention that the instances I shall give, with few exceptions, have been collected with little trouble, and without seeking, beyond my own small circle of acquaintance.

In the family of the above-named gentleman, Mr. R. E. S., there probably existed a faculty of presentiment; for, in the year 1810, his elder brother being Assistant-Surgeon on board the Gorgon, war-brig, his father dreamt that he was promoted to the Sparrow-hawk — a ship he had then never heard of; neither had the family received any intelligence of the young man for several months. He told his dream, and was well laughed at for his pains; but in a few weeks a letter arrived announcing the promotion.

When Lord Burghersh was giving theatrical parties at Florence, a lady, Mrs. M., whose presence was very important, excused herself one evening, being in great alarm from having dreamt in the night that her sister, in England, was dead, which proved to be the fact.

Mr. W., a young man at Glasgow College, not long since dreamt that his aunt in Russia was dead. He noted the date of his dream on the window-shutter of his chamber. In a short time the news of the lady's death arrived. The dates, however, did not accord; but, on mentioning the circumstance to a friend, he was reminded that the adherence of the Russians to the old style reconciled the difference.

A man of business, in Glasgow, lately

dreamt that he saw a coffin, on which was inscribed the name of a friend, with the date of his death. Some time afterwards he was summoned to attend the funeral of that person, who, at the time of the dream, was in good health, and he was struck with surprise on seeing the plate of the coffin bearing the very date he had seen in his dream.

A French gentleman, Monsieur de V., dreamt, some years since, that he saw a tomb, on which he read, very distinctly, the following date—23rd June, 184—; there were, also, some initials, but so much effaced that he could not make them out. He mentioned the circumstance to his wife; and, for some time, they could not help dreading the recurrence of the ominous month; but, as yean after year passed, and nothing happened, they had ceased to think of it, when, at last, the symbol was explained. On the 23rd of June, 1846, their only daughter died at the age of seventeen.

Thus far the instances I have related seem to resolve themselves into cases of simple clairvoyance, or second sight, in sleep, although, in using these words, I am very far from meaning to imply that I explain the thing, or unveil its mystery. The theory above alluded to, seems, as yet, the only one applicable to the facts, namely, that the senses, being placed in a negative and passive state, the universal sense of the immortal spirit within, which sees, and hears, and knows, or rather, in one word, *perceives*, without organs, becomes more or less free to work unclogged. That the soul is a mirror in which the spirit sees all things reflected, is a modification of this theory; but I confess I find myself unable to attach any idea to this latter form of expression. Another view, which I have heard suggested by an eminent person, is, that, if it be true, as maintained by Dr. Wigan, and some other physiologists, that our brains are double, it is possible that a polarity may exist between the two sides, by means of which the negative side may, under certian circumstances, become a mirror to the positive. It seems difficult to reconcile this notion with the fact, that these perceptions occur most frequently when the brain is asleep. How far the sleep is perfect and general, however, we can never know; and, of course, when the powers of speech and locomotion continue to be exercised, we are aware that it is only partial, in a more or less degree. In the case of magnetic sleepers, observation shows us, that the auditory nerves are aroused by being addressed, and fall asleep again as soon as they are left undisturbed. In most cases of natural sleep, the same process, if the voice were heard at all, would disperse sleep altogether; and it must be remembered that, as Dr. Holland says, sleep is a fluctuating condition, varying from one moment to another, and this allowance must be made when considering magnetic sleep also.

It is by this theory of the duality of the brain, which seems to have many arguments in its favour, and the alternate sleeping and waking of the two sides, that Dr. Wigan seeks to account for the state of double or alternate consciousness above alluded to; and also, for that strange sensation which most people have experienced, of having witnessed a scene, or heard a conversation, at some indefinite period before, or even in some earlier state of existence. He thinks that one-half of the brain being in a more active condition than the other, it takes cognizance of the scene first; and that thus the perceptions of the second, when they take place, appear to be a repetition of some former experiences. I confess this theory, as regards this latter phenomenon, is to me eminently unsatisfactory, and it is especially defective in not accounting for one of the most curious particulars connected with it, namely, that on these occasion, people not only seem to recognise the circumstances as having been experienced before; but they have, very frequently, an actual foreknowledge of what will be next said or done.

Now, the explanation of this mystery, I incline to think, may possibly lie in the hypothesis I have suggested; namely, that in profound, and what appears to us generally to have been dreamless sleep, we are clear-seers. The map of coming events lies open before us, the spirit surveys it; but with the awaking of the sensuous organs, this dream-life, with its aerial excursions, passes away; and we are translated into our other sphere of existence. But, occasionally, some flash of recollection, some ray of light, from this visionary world, in which we have been living, breaks in upon our external objective existence, and we recognize the locality, the voice, the very words, as being but a re-acting of some foregone scenes of a drama.

The faculty of presentiment, of which everybody must have heard instances, seems to have some affinity to the phenomenon last referred to. I am acquainted with a lady, in whom this faculty is in some degree developed, who has evinced it by a consciousness of the moment when a death was taking place in her family, or amongst her connexions, although she does not know who it is that has departed. I have heard of several cases of people hurrying home from a presentiment of fire; and Mr. M. of Calderwood was once, when absent from home, siezed with such an anxiety about his family, that, without being able in any way to account for it, he felt himself impelled to fly to them and remove them from the house they were inhabiting; one wing of which fell down immediately afterwards. No notion of such a misfortune had ever before occurred to him, nor was there any reason whatever to expect it; the accident originating from some defect in the foundations.

A circumstance, exactly similar to this, is related by Stilling, of Professor Böhm, teacher of Mathematics at Marburg; who being one evening in company, was suddenly seized with a conviction that he ought to go home. As, however, he was very comfortably taking his tea, and had nothing to do at home, he resisted the admonition; but it returned with such force that at length he was obliged to yield. On reaching his house, he found everything as he had left it; but he now felt himself urged to remove his bed from the corner in which it stood to another; but as it had always stood there, he resisted this impulsion also. However, the resistance was vain, absurd as it seemed, he felt he must do it; so he summoned the maid, and, with her aid, drew the bed to the other side of the room; after which he felt quite at ease and returned to spend the rest of the evening with his friends. At ten o'clock the party broke up, and he retired home and went to bed and to sleep In the middle of the night, he was awakened by a loud crash, and on looking out, he saw that a large beam had fallen, bringing part of the ceiling with it, and was lying exactly on the spot his bed had occupied.

A young servant girl in this neighbourhood, who had been several years in an excellent situation, where she was much esteemed, was suddenly seized with a presentiment that she was wanted at home; and, in spite of all representations, she resigned her place and set out on her journey thither; where, when she arrived, she found her parents extremely ill, one of them mortally, and in the greatest need of her seheces. No intelligence of their illness had reached her, nor could she herself in any way account for the impulse. I have heard of numerous well authenticated cases of people escaping drowning from being seized with an unaccountable presentiment of evil when there were no external signs whatever to justify the apprehension. The story of Cazotte as related by La Harpe is a very remarkable instance of this sort of faculty; and seems to indicate a power resembling that possessed by Zschokke, who relates of himself, in his autobiography, that, frequently whilst conversing with a stranger, the whole circumstances of that person's previous life were revealed to him, even comprising details of places and persons. In the case of Cazotte, it was the future that was laid open to him, and he foretold, to a company of eminent persons, in the year 1788, the fate which awaited each individual, himself included, in consequence of the revolution then commencing. As this story is already in print, I forbear to relate it.

One of the most remarkable cases of presentiment I know, is, that which occurred, not very long since, on board one of her Majesty's ships, when lying off Portsmouth. The officers being one day at the mess-table, a young Lieutenant P. suddenly laid down his knife and fork, pushed away his plate, and turned extremely pale. He then rose from the table, covering his face with his hands, and retired from the room The president of the mess, supposing him to be ill, sent one of the young men to enquire what was the matter. At first, Mr. P. was unwilling to speak; but on being pressed, he confessed that he had been seized by a sudden and irresistible impression, that a brother he had then in India was dead. "He died," said he, "on the 12th of August, at six o'clock; I am perfectly certain of it!" No arguments could overthrow this conviction, which, in due course of post, was verified to the letter. The young man had died at Cawnpore, at the precise period mentioned.

When any exhibition of this sort of faculty occurs in animals, which is by no means unfrequent, it is termed *instinct;* and we look upon it, as what it probably is, only another and more rare development of that intuitive knowledge which enables them to seek their food, and perform the other functions necessary to the maintenance of their existence, and the continuance of their race. Now, it is remarkable, that the life of an animal is a sort of dream-life; their ganglionic system is more developed than that of man, and the cerebral, less; and since it is doubtless, from the greater development of the ganglionic system in women, that they exhibit more frequent instances of such abnormal phenomena as I am treating of, than men, we may be, perhaps, justified in considering the faculty of presentiment in a human being as a suddenly awakened instinct; just as in an animal, it is an intensified instinct.

Everybody has either witnessed or heard of instances of this sort of presentiment, in dogs especially. For the authenticity of the following anecdote I can vouch; the traditions being very carefully preserved in the family concerned, from whom I have it. In the last century, Mr. P., a member of this family, who had involved himself in some of the stormy affairs of this northern part of the island, was one day surprised by seeing a favourite dog, that was lying at his feet, start suddenly up and seize him by the knee, which he pulled—not with violence, but in a manner that indicated a wish that his master should follow him to the door. The gentleman resisted the invitation for some time; till at length the perseverance of the animal arousing his curiosity, he yielded, and was thus conducted by the dog into the most sequestered part of a neighbouring thicket, where, however, he could see nothing to account for his dumb friend's proceeding, who now lay himself down, quite satisfied, and seemed to wish his master to follow the example; which, determined to pursue the adventure and find out, if possible, what was meant, he did. A considerable time now elapsed before the dog would consent to his master's going home; but at length he arose and led the way thither, when the first news Mr. P. heard was, that a party of soldiers had been there in quest of him; and he was shown the marks of their spikes, which had been thrust through the bed-clothes in their search. He fled, and ultimately escaped; his life being thus preserved by his dog.

Some years ago, at Plymouth, I had a brown spaniel that regularly, with great delight, accompanied my son and his nurse in their morning's walk. One day, she came to complain to me that Tiger would not go out with them. Nobody could conceive the reason of so unusual a caprice; and, unfortunately, we did not yield to it, but forced him to go. In less than a quarter of an hour he was brought back, so torn to pieces, by a savage dog that had just come ashore from a foreign vessel, that it was found necessary to shoot him immediately.

CHAPTER V.

WARNINGS.

This comparison, betwixt the power of presentiment in a human being and the instincts of an animal, may be offensive to some people; but it must be admitted, that, as far as we can see, the manifestation is the same, whatever be the cause. Now, the body of an animal must be informed by an immaterial principle —let us call it soul or spirit, or anything else; for it is evident that their actions are not the mere result of organization; and all I mean to imply is, that this faculty of fore-seeing must be inherent in intelligent spirit, let it be lodged in what form of flesh it may; whilst, with regard to what instinct is, we are, in the meanwhile, in extreme ignorance. *Instinct* being a word which, like *Imagination*, everybody uses, and nobody understands.

Ennemoser and Schubert believe, that the instinct by which animals seek their food consists in polarity, but I have met with only two modern theories which pretend to explain the phenomena of presentiment; the one is, that the person is in a temporarily magnetic state, and that the presentiment is a kind of clairvoyance. That the faculty, like that of prophetic dreaming, is constitutional, and chiefly manifested in certain families, is well established; and the very unimportant events, such as visits, and so forth, on which it frequently exercises itself, forbid us to seek an explanation in a higher source. It seems, also, to be quite independent of the will of the subject, as it was in the case of Zschokke, who found himself thus let into the secrets of persons in whom he felt no manner of interest; whilst, where the knowledge might have been of use to him, he could not command it. The theory of one-half of the brain in a negative state, serving as a mirror to the other half, if admitted at all, may answer as well, or better, for these waking presentiments, than for clear-seeing in dreams. But, for my own part, I incline very much to the views of that school of philosophers who adopt the first and more spiritual theory, which seems to me, to offer fewer difficulties, whilst, as regards our present nature, and our future hopes, it is certainly more satisfactory. Once admitted that the body is but the temporary dwelling of an immaterial spirit, the machine through which, and by which, in its normal states, the spirit alone can manifest itself, I cannot see any great difficulty in conceiving that, in certain conditions of that body, their relations may be modified, and that the spirit may perceive, by its own inherent quality, without the aid of its material vehicle; and, as this condition of the body may arise from causes purely physical, we see at once why the revelations frequently regard such unimportant events.

Plutarch, in his dialogue betwixt Lamprius and Ammonius, observes, that if the Dæmons, or protecting spirits, that watch over mankind are disembodied souls, we ought not to doubt that those spirits, even when in the flesh, possessed the faculties they now enloy, since we have no reason to suppose that any new ones are conferred at the period of dissolution; for these faculties must be inherent, although temporarily obscured, and weak and ineffective in their manifestations. As it is not when the sun breaks from behind the clouds that he first begins to shine, so it is not when the soul issues from the body, as from a cloud that envelopes it, that it first attains the power of looking into the future.

But the events foreseen are not always unimportant, nor is the mode of the communication always of the same nature. I have mentioned above, some instances wherein danger was avoided, and there are many of the same kind recorded in various works; and it is the number of instances of this description, corroborated by the universal agreement of all somnambulists of a higher order, which has induced a considerable section of the German psychologists to adopt the doctrine of guardian spirits—a doctrine which has prevailed, more or less, in all ages; and has been considered by many theologians to be supported by the Bible. There is in this country, and I believe in France, also, though with more exceptions, such an extreme aversion to admit the possibility of anything like what is called supernatural agency, that the mere avowal of such a persuasion is enough to discredit one's understanding with a considerable part of the world; not excepting those who profess to believe in the scriptures. Yet, even apart from this latter authority, I cannot see anything repugnant to reason in such a belief. As far as we see of nature, there is a continued series from the lowest to the highest; and what right have we to conclude that we are the last link of the chain? Why may there not be a gamut of beings? That such should be the case, is certainly in accordance with all that we see; and that we do not see them, affords, as I have said above, not a shadow of argument against their existence; man, immersed in business and pleasure, living only his sensuous life, is too apt to forget how limited those senses are, how merely designed for a temporary purpose, and how much may exist of which they can take no cognizance.

The *possibility* admitted, the chief arguments against the *probability* of such a guardianship, are the interference it implies with the free-will of man, on the one hand, and the rarity of this interference, on the other. With respect to the first matter of free-will, it is a subject of acknowledged difficulty, and beyond the scope of my work. Nobody can honestly look back upon his past life without feeling perplexed by the question, of how far he was, or was not able, at the moment, to resist certain impulsions, which caused him to commit wrong or imprudent actions; and it must, I fear, ever remain a *quæstio vexata* how far our virtues and vices depend upon our organization; an organization whose constitution is beyond our own power, in the first instance, although we may certainly improve or deteriorate it; but which we must admit, at the same time, to be, in its present deteriorated form, the ill result of the world's corruption, and the inherited penalty of the vices of our predecessors; whereby the sins of the fathers are visited upon the children unto the third and fourth generation.

There is, as the Scriptures say, but one way to salvation, though there are many to perdition, that is, though there are many wrongs, there is only one right; for truth is one, and our true liberty consists in being free to follow it; for we cannot imagine that anybody seeks his own perdition, and nobody, I conceive, loves vice for its own sake, as others love virtue, that is, because it *is* vice; so that, when they follow its dictates, we must conclude that they are not free, but in bondage, whose ever bond-slave they be, whether of an evil spirit, or of their own organization; and, I think, every human being who looks into himself will feel, that he is, in effect then only *free* when he is obeying the dictates of virtue; and that the language of Scripture, which speaks of sin as a bondage, is not only metaphorically, but literally, true.

The warning a person of an impending danger, or error, implies no constraint; the subject of the warning is free to take the hint or not, as he pleases; we receive many cautions, both from other people and from our own consciences, which we refuse to benefit by.

With regard to the second objection, it seems to have greater weight; for although the instances of presentiment are very numerous taken apart, they are, certainly, as far as we know, still but exceptional cases. But here we must remember, that an influence of this sort might be very continuously, though somewhat remotely, excercised in favour of an individual, without the occurrence of any instance of so striking a nature, as to render the interference manifest; and certain it is, that some people—I have met with several—and very sensible persons, too, have all their lives an intuitive persuasion of such a guardianship existing in relation to themselves. That in our normal states it was not intended we should hold sensible communion with the invisible world, seems evident; but nature abounds in exceptions; and there may be conditions regarding both parties, the incorporated and the unincorporated spirit, which may at times bring them into a more intimate relation. No one who believes that consciousness is to survive the death of the body, can doubt that the released spirit will then hold communion with its congeners; it being the fleshly tabernacles we inhabit which alone disables us from doing so at present; but since the constitutions of bodies vary exceedingly, not only in different individuals, but in the same individuals at different times, may we not conceive the possibility of there existing conditions, which by diminishing the obstructions, render this communion practicable within certain limits? For there, certainly, are recorded and authentic instances of presentiments and warnings, that with difficulty admit of any other explanation; and that these admonitions are more frequently received in the state of sleep than of vigilance, rather furnishes an additional argument in favour of the last hypothesis; for if there be any foundation for the theories above suggested, it is then, that the sensuous functions being in abeyance, and the external life thereby shut out from us, the spirit would be most suscep tible to the operations of spirit, whether of our deceased friends or of appointed ministers, if such there be. Jung Stelling is of opinion that we must decide from the aim and object of the revelation, whether it be a mere development of the faculty of presentiment, or a case of spiritual intervention; but this would surely be a very erroneous mode of judging, since the presentiment that foresees a visit, may foresee a danger, and show us how to avoid it, as in the following instance:—

A few years ago, Dr. W., now residing at Glasgow, dreamt that he received a summons to attend a patient at a place some miles from where he was living; that he started on horseback, and that as he was crossing a moor, he saw a bull making furiously at him, whose horns he only escaped by taking refuge on a spot inaccessible to the animal; where he waited a long time, till some people, observing his situation, came to his assistance and released him. Whilst at breakfast, on the following morning, the summons came; and, smiling at the odd *coincidence*, he started on horseback. He was quite ignorant of the road he had to go; but, by and by, he arrived at the moor, which he recognised, and presently the bull appeared, coming full tilt towards him. But his dream had shown him the place of refuge, for which he instantly made; and there he spent three or four bours, besieged by the animal, till the country people set him free. Dr. W. declares, that but for the dream, he should not have known in what direction to runt for safety.

A Butcher named Bone, residing at Holytown, dreamt a few years since, that he was stopt at a particular spot on his way to market, whither he was going on the following day to purchase cattle, by two men in blue clothes, who cut his throat. He told the dream to his wife, who laughed at him; but as it was repeated two or tnree times, and she saw he was really alarmed, she advised him to join somebody who was going the same road. He accordingly listened till he heard a cart passing his door, and then went out and joined the man, telling him the reason for so doing. When they came to the spot, there actually stood the two men in blue clothes, who, seeing he was not alone, took to their heels and ran.

Now, although the dream was here probably the meansof saving Bone's life, there is no reason to suppose this a case of what is called *supernatural intervention*. The phenomenon would be sufficiently accounted for by the admission of the hypothesis I have suggested; namely, that he was aware of the impending danger in his sleep, and had been able, from some cause unknown to us, to convey the recollection into his waking state.

I know instances in which, for several mornings previous to the occurrence of a calamity, persons have awakened with a painful sense of misfortune, for which they could not account, and which was dispersed as soon as they had time to reflect that they had no cause for uneasiness. This is the only kind of presentiment I ever experienced myself; but it has occurred to me twice, in a very marked and unmistakeable manner. As soon as the intellectual life, the life of the brain, and the external world broke in, the instinctive life receded, and the intuitive knowledge was obscured. Or, according to Dr. Ennemoser's theory, the polar relations changed, and the nerves were busied with conveying sensuous impressions to the brain, their sensibility or positive state now being transferred from the internal to the external periphery. It is by the contrary change that Dr. Ennemoser seeks to explain the insensibility to pain of mesmerised patients.

A circumstance of a similar kind to the above occurred in a well known family in Scotland, the Rutherfords of E.—A lady dreamt that her aunt, who resided at some distance, was murdered by a black servant.

Impressed with the liveliness of the vision, she could not resist going to the house of her relation, where the man she had dreamt of, whom I think she had never before seen, opened the door to her. Upon this, she induced a gentleman to watch in the adjoining room during the night; and towards morning hearing a foot upon the stairs, he opened the door and discovered the black servant carrying up a coal scuttle full of coals, for the purpose, as he said, of lighting his mistress's fire. As this motive did not seem very probable, the coals were examined and a knife found hidden amongst them, with which, he afterwards confessed, he intended to have murdered his mistress, provided she made any resistance to a design he had formed, of robbing her of a large sum of money, which he was aware she had that day received.

The following case has been quoted in several medical works—at least in works written by learned doctors, and on that account I should not mention it here, but for the purpose of remarking on the extraordinary facility with which, whilst they do not question the fact, they dispose of the mystery.

Mr. D., of Cumberland, when a youth, came to Edinburgh, for the purpose of attending College, and was placed under the care of his uncle and aunt, Major and Mrs. Griffiths, who then resided in the castle. When the fine weather came, the young man was in the habit of making frequent excursions, with others of his own age and pursuits; and one afternoon he mentioned that they had formed a fishing party, and had bespoken a boat for the ensuing day. No objections were made to this plan; but in the middle of the night, Mrs. Griffiths screamed out, "The boat is sinking! Oh, save them!" Her husband said, he supposed she had been thinking of the fishing party; but she declared she had never thought about it, at all, and soon fell asleep again. But, ere long, she awoke a second time, crying out that she "saw the boat sinking!" "It must have been the remains of the impression made by the other dream," she suggested to her husband, "for I have no uneasiness, whatever, about the fishing party."—but on going to sleep once more, her husband was again disturbed by her cries, "they are gone!" she said, "the boat has sunk!" She now, really, became alarmed, and, without waiting for morning, she threw on her dressing gown, and went to Mr. D., who was still in bed, and, whom with much difficulty, she persuaded to relinquish his proposed excursion. He, consequently, sent his servant to Leith, with an excuse; and the party embarked without him. The day was extremely fine, when they put to sea; but some hours afterwards, a storm arose, in which the boat foundered; nor did any one of the number survive to tell the tale.

"This dream is easily accounted for," say the learned gentlemen above alluded to, "from the dread all women have of the water, and the danger that attends boating on the Frith of Forth!" Now, I deny that all women have a dread of the water, and there is not the slightest reason for concluding that Mrs. Griffiths had. At all events, she affirms that she felt no uneasiness at all about the party, and one might take leave to think that her testimony upon that subject is of more value than that of persons who never had any acquaintance with her, and who were not so much as born at the time the circumstance occurred, which was in the year 1731. Besides, if Mrs. Griffith's dream arose simply from "the dread all women have of the water,' and that its subsequent verification was a mere coincidence, since women constantly risk their persons for voyages, and boating excursions, such dreams should be extremely frequent; the fact of there being any accident impending, or not, having, according to this theory, no relation whatever to the phenomenon: And as for the danger that attends boating on the Frith of Forth, we must naturally suppose that had it been considered so imminent, Major Griffiths would have, at least, endeavoured to dissuade a youth that was placed under his protection from risking his life so imprudently. It would be equally reasonable to explain away Dr. W.'s dream, by saying, that all gentlemen who have to ride across commons are in great dread of encountering a bull—commons, in general, being infested by that animal.

Miss D., a friend of mine, was some time since invited to join a pic-nic excursion into the country. Two nights before the day fixed for the expedition, she dreamt that the carriage she was to go in, was overturned down a precipice. Impressed with her dream, she declined the excursion, confessing her reason, and advising the rest of the party to relinquish their project. They laughed at her, and persisted in their scheme. When, subsequently, she went to enquire how they had spent the day, she found the ladies confined to their beds, from injuries received; the carriage haivng been overturned down a precipice. Still, this was only a coincidence!

Another specimen of the haste with which people are willing to dispose of what they do not understand, is afforded by a case that occurred, not many years since, in the north of Scotland, where a murder having been committed, a man came forward saying that he had dreamt that the pack of the murdered pedlar was hidden in a certain spot; where, on a search being made, it was actually found. They at first concluded he was himself the assassin, but the real criminal was afterwards discovered; and it being asserted, though I have been told erroneously, that the two men had passed some time together, since the murder, in a state of intoxication, it was decided that the crime and the place of concealment had been communicated to the pretended dreamer; and all who thought otherwise were laughed at; for why, say the rationalists, should not Providence have so ordered the dream as to have prevented the murder altogether?

Who can answer that question, and whither would such a discussion lead us? Moreover, if this faculty of presentiment be a natural one, though only imperfectly and capriciously developed, there may have been no design in the matter; it is an accident, just in the same sense as an illness is an accident; that is, not without cause, but without a cause that we can penetrate. If, on the other hand, we have recourse to the intervention of spiritual beings, it may be answered that we are entirely ignorant of the conditions under which any such communication is possible; and that we cannot therefore come to any conclusions as to why so much is done, and no more.

But there is another circumstance to be observed in considering the case, which is, that the dreamer is said to have passed some days in a state of intoxication. Now, even supposing this had been true, it is well-known that the excitement of the brain, caused by intoxication, has occasionally produced a very remarkable exaltation of certain faculties. It is by means, either of intoxicating draughts or vapours, that the soothsayers of Lapland and Siberia place themselves in a condition to vaticinate;

and we have every reason to believe that drugs, producing similar effects, were resorted to by the thaumaturgists of old, and by the witches of later days, of which I shall have more to say hereafter. But as a case in point, I may here allude to the phenomena exhibited in a late instance of the application of ether, by Professor Simpson, of Edinburgh, to a lady who was at the moment under circumstances not usually found very agreeable. She said that she was amusing herself delightfully by playing over a set of quadrilles, which she had known in her youth, but had long forgotten; but she now perfectly remembered them, and had played them over several times. Here was an instance of the exaltation of a faculty from intoxication, similar to that of the woman who, in her delirium, spoke a language which she had only heard in her childhood, and of which, in her normal state, she had no recollection.

That the inefficiency of the communication, or presentiment, or whatever it may be, is no argument against the fact of such dreams occurring, I can safely assert, from cases which have come under my own knowledge. A professional gentleman, whose name would be a warrant for the truth of whatever he relates, told me the following circumstance regarding himself. He was, not very long since, at the sea-side, with his family, and, amongst the rest, he had with him one of his sons, a boy about twelve years of age, who was in the habit of bathing daily, his father accompanying him to the water side. This practice had continued during the whole of their visit, and no idea of danger or accident had ever occurred to anybody. On the day preceding the one appointed for their departure, Mr. H., the gentleman in question, felt himself, after breakfast, surprised by an unusual drowsiness, which he, having vainly struggled to overcome, at length fell asleep in his chair, and dreamt that he was attending his son to the bath as usual, when he suddenly saw the boy drowning, and that he himself had rushed into the water, dressed as he was, and brought him ashore. Though he was quite conscious of the dream when he awoke, he attached no importance to it; he considered it merely a dream, no more; and when, some hours afterwards, the boy came into the room, and said, "Now, papa, it's time to go; this will be my last bath;" his morning's vision did not even recur to him. They walked down to the sea, as usual, and the boy went into the water, whilst the father stood composedly watching him from the beach, when, suddenly, the child lost his footing, a wave had caught him, and the danger of his being carried away was so imminent, that, without even waiting to take off his great coat, boots, or hat, Mr. H. rushed into the water, and was only just in time to save him.

Here is a case of undoubted authenticity, which I take to be an instance of clear-seeing, or second sight, in sleep. The spirit, with its intuitive faculty, saw what was impending; the sleeper remembered his dream, but the intellect did not accept the warning; and, whether that warning was merely a subjective process—the clear-seeing of the spirit—or whether it was effected by any external agency, the free will of the person concerned was not interfered with.

I quote the ensuing similar case from the Frankfort Journal, 25th June, 1837:—"A singular circumstance is said to be connected with the late attempt on the life of the Archbishop of Autun. The two nights preceding the attack, the prelate dreamt that he saw a man, who was making repeated efforts to take away his life, and he awoke in extreme terror and agitation from the exertions he had made to escape the danger. The features and appearance of the man were so clearly imprinted on his memory, that he recognized him the moment his eye fell upon him, which happened as he was coming out of church. The bishop hid his face, and called his attendants, but the man had fired before he could make known his apprehensions. Facts of this description are far from uncommon. It appears that the assassin had entertained designs against the lives of the Bishops of Dijon, Burgos, and Nevers."

The following case, which occurred a few years since, in the North of England, and which, I have from the best authority, is remarkable from the inexorable fatality which brought about the fulfilment of the dream:—Mrs. K., a lady of family and fortune in Yorkshire, said to her son, one morning, on descending to breakfast, "Henry, what are you going to do to-day?"

"I am going to hunt," replied the young man."

"I am very glad of it," she answered. "I should not like you to go shooting, for I dreamt last night that you did so, and were shot. The son answered gaily, that he would take care not to be shot, and the hunting party rode away; but, in the middle of the day they returned, not having found any port Mr.

B., a visitor in the house, then proposed that they should go out with their guns, and try to find some woodcocks. "I will go with you," returned the young man, "but I must not shoot to-day, myself, for my mother dreamt last night I was shot; and, although it is but a dream, she would be uneasy."

They went, Mr. B. with his gun, and Mr. K. without; but shortly afterwards the beloved son was brought home dead. A charge from the gun of his companion had struck him in the eye, entered his brain, and killed him on the spot. Mr. B., the unfortunate cause of this accident, and also the narrator of it, died but a few weeks since.

It is well known that the murder of Mr. Percival, by Bellingham, was seen in sleep by a gentleman at York, who actually went to London in consequence of his dream, which was several times repeated. He arrived too late to prevent the calamity; neither would he have been believed, had he arrived earlier.

In the year 1461, a merchant was travelling towards Rome, by Sienna, when he dreamt that his throat was cut. He communicated his dream to the host of the inn, who did not like it, and advised him to pray and confess. He did so, and then rode forth, and was presently attacked by the priest he had confessed to, who had thus learnt his apprehensions. He killed the merchant, but was betrayed, and disappointed of his gains, by the horse taking fright, and running back to the inn with the money bags.

I have related this story, though not a new one, on account of its singular resemblance to the following, which I take from a newspaper paragraph; but which I find mentioned as a fact in a continental publication :—

"SINGULAR VERIFICATION OF A DREAM.—A letter from Hamburgh, contains the following curious story, relative to the verification of a dream. It appears that a locksmith's apprentice, one morning lately, informed his master (Claude Soller), that on the previous night he dreamt that he had been assassinated on the road to Bergsdorff, a little town at about two hours' distance from Hamburgh. The master laughed at the young man's credulity, and to prove that he himself had little faith in dreams, insisted upon sending him to Bergsdorff, with 140 rix dollars (£22 8s.), which he owed to his brother-in-law, who resided in the town. The apprentice, after in vain imploring his master to change his intention, was compelled to set out, at about eleven o'clock. On arriving at the village of Bill-waerder, about half-way between Hamburgh and Bergsdorff, he recollected his dream with terror, but perceiving the baillie of the village at a little distance, talking to some of his workmen, he accosted him, and acquainted him with his singular dream, at the same time requesting, that, as he had money about his person, one of his workmen might be allowed to accompany him for protection across a small wood which lay in his way. The baillie smiled, and, in obedience to his orders, one of his men set out with the young apprentice. The next day, the corpse of the latter was conveyed by some peasants to the baillie, along with a reaping-hook, which had been found by his side, and, with which, the throat of the murdered youth had been cut. The baillie immediately recognised the instrument as one which he had on the previous day given to the workman who had served as the apprentice's guide, for the purpose of pruning some willows. The workman was apprehended, and, on being confronted with the body of his victim, made a full confession of his crime, adding, that the recital of the dream had alone prompted him to commit the horrible act. The assassin, who is thirty-five years of age, is a native of Bill-waerder, and, previously to the perpetration of the murder, had always borne an irreproachable character."

The life of the great Harvey was saved by the Governor of Dover refusing to allow him to embark for the Continent, with his friends. The vessel was lost, with all on board; and the Governor confessed to him, that he had detained him in consequence of an injunction he had received in a dream to do so.

There is a very curious circumstance related by Mr. Ward, in his "Illustrations of Human Life," regarding the late Sir Evan Nepean, which, I believe is perfectly authentic. I have, at least been assured, by persons well acquainted with him, that he himself testified to its truth.

Being, at the time, Secretary to the Admiraly, he found himself one night unable to sleep, and urged by an undefinable feeling that he must rise, though it was then only two o'clock. He accordingly did so, and went into the park, and from that to the Home Office, which he entered by a private door, of which he had the key. He had no object in doing this, and to pass the time, he took up a newspaper that was lying on the table, and there read a paragraph to the effect, that a reprieve had been dispatched to York, for the men condemned for coining.

The question occurred to him, was it indeed dispatched? He examined the books and found it was not; and it was only by the most energetic proceedings that the thing was carried through, and reached York in time to save the men.

Is not this like the agency of a protecting spirit, urging Sir Evan to this discovery, in order that these men might be spared; or that those concerned might escape the remorse they would have suffered for their criminal neglect?

It is a remarkable fact, that somnambules of the highest order believe themselves attended by a protecting spirit. To those who do not believe, because they have never witnessed the phenomena of somnambulism, or who look upon the disclosures of persons in that state, as the mere raving of hallucination, this authority will necessarily have no weight; but even to such persons, the universal coincidence, must be considered worthy of observation, though it be regarded only as a symptom of disease. I believe I have remarked, elsewhere, that many persons, who have not the least tendency to somnambulism, or any proximate malady, have, all their lives, an intuitive feeling of such a guardianship; and, not to mention Socrates and the ancients, there are, besides, numerous recorded cases in modern times, in which persons, not somnambulic, have declared themselves to have seen and held communication with their spiritual protector.

The case of the girl called Ludwiger, who, in her infancy, had lost her speech, and the use of her limbs, and who was earnestly committed by her mother, when dying, to the care of her elder sisters, is known to many. These young women piously fulfilled their engagement, till the wedding-day of one of them caused them to forget their charge. On recollecting it, at length, they hastened home, and found the girl to their amazement, sitting up in her bed, and she told them, that her mother had been there and given her food. She never spoke again, and soon after died. This circumstance occurred at Dessau, not many years since; and is, according to Schubert, a perfectly established fact in that neighbourhood. The girl, at no other period of her life exhibited any similar phenomena, nor had she eyer displayed any tendency to spectral illusions.

The wife of a respectable citizen, named Arnold, at Heilbronn, held constant communications with her protecting spirit, who warned her of impending dangers, approaching visitors, and so forth. He was only once visible to her, and it was in the form of an old man; but his presence was felt by others as well as herself, and they were sensible that the air was stirred, as by a breath.

Jung Stilling publishes a similar account, which was bequeathed to him by a very worthy and pious minister of the church. The subject of the guardianship was his own wife; and the spirit first appeared to her after her marriage, in the year 1799, as a child, attired in a white robe, whilst she was busy in her bed-chamber. She stretched out her hand to take hold of the figure, but it disappeared. It frequently visited her afterwards, and in answer to her enquiries, it said, "I died in my childhood!" It came to her at all hours, whether alone or in company, and not only at home, but elsewhere, and even when travelling, assisting her when in danger; it sometimes floated in the air, spake to her in its own language, which, somehow, she says, she understood, and could speak, too; and it was once seen by another person. He bade her call him

Immanuel. She earnestly begged him to show himself to her husband, but he alleged that it would make him ill, and cause his death. On asking him *wherefore*, he answered, "few persons are able to see such things."

Her two children, one six years old, and the other younger, saw this figure as well as herself.

Schubert, in his "Geschichte der Seele," relates that the ecclesiastical councillor Schwartz, of Heidelberg, when about twelve years of age, and at a time that he was learning the Greek language, but knew very little about it, dreamt that his grandmother, a very pious woman, to whom he had been much attached, appeared to him, and unfolded a parchment, inscribed with Greek characters, which foretold the fortunes of his future life. He read it off with as much facility as if it had been in German; but being dissatisfied with some particulars of the prediction, he begged they might be changed. His grandmother answered him in Greek, whereupon he awoke, remembering the dream, but, in spite of the efforts to arrest them, he was unable to recall the particulars the parchment had contained. The answer of his grandmother, however, he was able to grasp before it had fled his memory, and he wrote down the words; but the meaning of them he could not discover, without the assistance of his Grammar and Lexicon. Being interpreted, they proved to be these—"As it is prophecied to me, so I prophecy to thee!" He had written the words in a volume of Gessner's works, being the first thing he laid his hand on; and he often philosophized on them in later days, when they chanced to meet his eye. How, he says, should he have been able to read and produce that in his sleep, which, in his waking state, he would have been quite incapable of? "Even long after, when I left school," he adds, "I could scarcely have put together such a sentence; and it is extremely remarkable that the feminine form was observed in conformity with the sex of the speaker. The words were these — *Ταῦτα χρησμῳδηθεισα Χρησμωδέ'ω σοι.*

Grotius relates, that when Mr. de Saumaise was councillor of the Parliament at Dijon, a person who knew not a word of Greek, brought him a paper, on which was written some words in that language, but not in the character. He said that a voice had uttered them to him in the night, and that he had written them down, imitating the sound as well as he could.

Mons. de Saumaise made out that the signification of the words, was; "Begone! do you not see that death impends?" Without comprehending what danger was predicted, the person obeyed the mandate and departed. On that night the house that he had been lodging in, fell to the ground.

The difficulty in these two cases is equally great, apply to it whatever explanation we may; for even if the admonitions proceeded from some friendly guardian, as we might be inclined to conclude, it is not easy to conceive why they should have been communicated in a language the persons did not understand.

After the death of Dante, it was discovered that the thirteenth canto of the "Paradiso" was missing; great search was made for it, but in vain; and to the regret of everybody concerned, it was at length concluded that it had either never been written, or had been destroyed. The quest was therefore given up, and some months had elapsed, when Pietro Allighieri, his son, dreamt that his father appeared to him and told him that if he removed a certain pannel near the window of the room, in which he had been accustomed to write, the thirteenth canto would be found. Pietro told his dream and was laughed at, of course; however, as the canto did not turn up, it was thought as well to examine the spot indicated in the dream. The pannel was removed, and there lay the missing canto behind it; much mildewed, but fortunately, still legible.

If it be true that the dead do return sometimes to solve our perplexities, here was not an unworthy occasion for the exercise of such a power. We can imagine the spirit of the great poet still clinging to the memory of his august work, immortal as himself—the record of those high thoughts which can never die.

There are numerous curious accounts extant of persons being awakened by the calling of a voice which announced some impending danger to them. Three boys are sleeping in the wing of a castle, and the eldest is awakened by what appears to him to be the voice of his father calling him by name. He rises and hastens to his parent's chamber, situated in another part of the building, where he finds his father asleep; who, on being awakened, assures him that he had not called him, and the boy returns to bed. But he is scarcely asleep, before the circumstance recurs, and he again goes to his father with the same result. A third time he falls asleep, and a third time he is aroused by the voice, too distinctly heard for him to doubt his senses; and now, alarmed at he knows not what, he rises and takes his brothers with him to his father's chamber; and whilst they are discussing the singularity of the circumstance, a crash is heard, and that wing of the castle in which the boys slept, falls to the ground. This incident excited so much attention in Germany that it was recorded in a ballad.

It is related by Amyraldus, that Monsieur Calignan, Chancellor of Navarre, dreamed three successive times in one night, at Berne, that a voice called to him and bade him quit the place, as the plague would soon break out in that town; that, in consequence, he removed his family, and the result justified his flight.

A German physician relates, that a patient of his told him, that he dreamt repeatedly, one night, that a voice bade him go to his hop-garden, as there were thieves there. He resisted the injunction some time, till, at length, he was told that, if he delayed any longer, he would lose all his produce. Thus urged, he went at last, and arrived just in time to see the thieves, loaded with sacks, making away from the opposite side of the hop-ground.

A Madame Von Militz, found herself under the necessity of parting with a property which had long been in her family. When the bargain was concluded, and she was preparing to remove, she solicited permission of the new proprietor to carry away with her some little relic as a memento of former days—a request which he uncivilly denied. On one of the nights that preceded her departure from the home of her ancestors, she dreamt that a voice spoke to her, and bade her go to the cellar and open a certain part of the wall, where she would find something that nobody would dispute with her. Impressed with her dream, she sent for a bricklayer, who, after long seeking, discovered a place which appeared less solid than the rest. A hole was made, and, in a niche, was found a goblet, which contained som-thing that looked like a pot pourri. On shaking out the contents, there lay at the bottom a small ring, on which was engraven the name *Anna Von Militz.*

A friend of mine, Mr. Charles Kirkpatrick Sharpe, has some coins that were found exactly in the same manner. The child of a Mr. Christison, in whose house his father was lodging, in the year 1781, dreamt that there was a treasure hid in the cellar. Her father had no faith in the dream; but Mr.

S. had the curiosity to have the place dug up, and a copper pot was found, full of coins.

A very singular circumstance was related to me lately, by Mr. J. J., as having occurred not long since to himself. A tonic had been prescribed to him by his physician, for some slight derangement of the system, and, as there was no good chemist in the village he inhabited, he was in the habit of walking to a town about five miles off, to get the bottle filled as occasion required. One night, that he had been to M. for this purpose, and had obtained his last supply, for he was now recovered, and about to discontinue the medicine, a voice seemed to warn him that some great danger was impending, his life was in jeopardy; then he heard, but not with his outward ear, a beautiful prayer. "It was not myself that prayed," he said, "the prayer was far beyond anything I am capable of composing—it spoke of me in the third person, always as *he;* and supplicated that for the sake of my widowed mother this calamity might be averted. My father had been dead some months. I was sensible of all this, yet I cannot say whether I was asleep or awake. When I rose in the morning, the whole was present to my mind, although I had slept soundly in the interval;

I felt, however, as if there was some mitigation of the calamity, though what the danger was with which I was threatened, I had no notion. When I was dressed, I prepared to take my medicine, but, on lifting the bottle, I fancied that the colour was not the same as usual. I looked again, and hesitated, and finally, instead of taking two table spoonfuls, which was my accustomed dose, I took but one. Fortunate it was that I did so ; the apothecary had made a mistake; the drug was poison ; I was seized with a violent vomiting, and other alarming symptoms, from which I with difficulty recovered. Had I taken the two spoonfuls, I should, probably, not have survived to tell the tale."

The manner in which I happened to obtain these particulars is not uninteresting. I was spending the evening with Mr. Wordsworth, at Ridal, when he mentioned to me that a stranger, who had called on him that morning, had quoted two lines from his poem of "Laodamia," which, he said, to him had a peculiar interest. They were these : —

> "The invisible world with thee hath sympathised;
> Be thy affections raised and solemnised."

"I do not know what he alludes to," said Mr. Wordsworth ; "but he gave me to understand that these lines had a deep meaning for him, and that he had himself been the subject of such a sympathy."

Upon this, I sought the stranger, whose address the poet gave me, and thus learnt the above particulars from himself. His very natural persuasion was, that the interceding spirit was his father. He described the prayer as one of earnest anguish.

One of the most remarkable instances of warning that has come to my knowledge, is that of Mr. M., of Kingsborough. This gentleman, being on a voyage to America, dreamt, one night, that a little old man came into his cabin and said, "Get up! Your life is in danger!" Upon which, Mr. M. awoke; but considering it to be only a dream, he soon composed himself to sleep again. The dream, however, if such it were, recurred, and the old man urged him still more strongly to get up directly; but he still persuaded himself it was only a dream; and after listening a few minutes, and hearing nothing to alarm him, he turned round and addressed himself once more to sleep. But now the old man appeared again, and angrily bade him rise instantly, and take his gun and ammunition with him, for he had not a moment to lose. The injunction was now so distinct that Mr. M. felt he could no longer resist it; so he hastily dressed himself, took his gun, and ascended to the deck, where he had scarcely arrived, when the ship struck on a rock, which he and several others contrived to reach. The place, however, was uninhabited, and, but for his gun, they would never have been able to provide themselves with food till a vessel arrived to their relief.

Now these can scarcely be looked upon as instances of clear seeing, or of second sight in sleep, which, in Denmark, is called *first-seeing*, I believe; for in neither case did the sleeper perceive the danger, much less the nature of it. If, therefore, we refuse to attribute them to some external protecting influence, they resolve themselves into cases of vague presentiment; but it must then be admitted that the mode of the manifestation is very extraordinary; so extraordinary, indeed, that we fall into fully as great a difficulty as that offered by the supposition of a guardian spirit.

An American clergyman told me that an old woman, with whom he was acquainted, who had two sons, heard a voice say to her in the night, "John's dead!" This was her eldest son. Shortly afterwards, the news of his death arriving, she said to the person who communicated the intelligence to her, "If John's dead, then I know that David is dead, too, for the same voice has since told me so;" and the event proved that the information, whenceever it came, was correct.

Not many years since, Captain S. was passing a night at the Manse of Strachur, in Argyleshire, then occupied by a relation of his own; shortly after he had lain down in bed, the curtains were opened, and somebody looked in upon him. Supposing it to be some inmate of the house, who was not aware that the bed was occupied, he took no notice of the circumstance, till, it being two or three times repeated, he at length said, "What do you want? Why do you disturb me in this manner?"

"I come," replied a voice, "to tell you, that this day twelvemonth you will be with your father!"

After this, Captain S. was no more disturbed. In the morning, he related the circumstance to his host; but, being an entire disbeliever in all such phenomena, without attaching any importance to the warning.

In the natural course of events, and quite irrespective of this visitation, on that day twelvemonth he was again at the Manse of Strachur, on his way to the North, for which purpose it was necessary that he should cross the ferry to Craigie. The day was, however, so exceedingly stormy, that his friend begged him not to go; but he pleaded his business, adding that he was determined not to be withheld from his intention by the ghost, and, although the minister delayed his departure, by engaging him in a game of backgammon, he at length started up, declaring he could stay no longer. They, therefore, proceeded to the water, but they found the boat moored to the side of the lake, and the boatman assured them that it would be impossible to cross. Captain S., however, insisted, and, as the old man was firm in his refusal, he became somewhat irritated, and laid his cane lightly across his shoulders.

"It ill becomes you, sir," said the ferryman, "to strike an old man like me; but, since you will have your way, you must; I cannot go with you, but my son will; but you will never reach the other side; he will be drowned, and you too."

The boat was then set afloat, and Captain S., together with his horse and servant, and the ferryman's son, embarked in it.

The distance was not great, but the storm was tremendous ; and, after having with great difficulty got half way across the lake, it was found impossible to proceed. The danger of tacking, was, of course, considerable ; but, since they could not advance, there was no alternative but to turn back, and it was resolved to attempt it. The manœuvre, however, failed ; the boat capsized, and they were all precipitated into the water.

" You keep hold of the horse, I can swim," said Captain S. to his servant, when he saw what was about to happen.

Being an excellent swimmer, and the distance from the shore inconsiderable, he hoped to save himself, but he had on a heavy top coat, with boots and spurs. The coat he contrived to take off in the water, and then struck out with confidence ; but, alas, the coat had got entangled with one of the spurs, and, as he swam, it clung to him, getting heavier and heavier, as it became saturated with water, ever dragging him beneath the stream. He, however, reached the shore, where his anxious friend still stood watching the event, and, as the latter bent over him, he was just able to make a gesture with his hand, which seemed to say, "You see, it was to be!" and then expired.

The boatman was also drowned; but, by the aid of the horse, the servant escaped.

As I do not wish to startle my readers nor draw too suddenly on their faith, I have commenced with this class of phenomena, which it must be admitted are sufficiently strange, and, if true, must also be admitted to be well worthy of attention. No doubt, these cases, and still more those to which I shall next proceed, give a painful shock to the received notions of polished and educated society in general; especially in this country, where the analytical or scientifical psychology of the eighteenth century has almost entirely superseded the study of synthetic or philosophical psychology. It has become a custom to look at all the phenomena regarding man in a purely physiological point of view; for although it is admitted that he has a mind, and although there is such a science as metaphysics, the existence of what we call mind, is never considered but as connected with the body. We know that body can exist without mind; for, not to speak of certain living conditions, the body subsists without mind when the spirit has fled; albeit, without the living principle it can subsist but for a short period, except under particular circumstances; but we seem to have forgotten that mind, though very dependant upon body as long as the connexion between them continues, can yet subsist without it. There have, indeed, been philosophers, purely materialistic, who have denied this; but they are not many; and not only the whole Christian world, but all who believe in a future state, must perforce admit it; for even those who hold that most unsatisfactory doctrine, that there will be neither memory nor consciousness till a second incorporation takes place, will not deny that the mind, however in a state of abeyance and unable to manifest itself, must still subsist, as an inherent property of man's immortal part. Even if, as some philosophers believe, the spirit, when freed from the body by death, returns to the Deity and is re-absorbed in the being of God, not to become again a separate entity until re-incorporated, still, what we call mind cannot be disunited from it. And when once we have begun to conceive of mind, and consequently of perception, as separated from and independent of bodily organs, it will not be very difficult to apprehend that those bodily organs must circumscribe and limit the view of the spiritual in-dweller, which must otherwise be necessarily perceptive of spirit like itself, though perhaps unperceptive of material objects and obstructions.

"It is perfectly evident to me," said Socrates, in his last moments, "that, to see clearly, we must detach ourselves from the body, and perceive by the soul alone. Not whilst we live, but when we die, will that wisdom which we desire and love, be first revealed to us; it must be then, or never, that we shall attain to true understanding and knowledge; since by means of the body we never can. But if, during life, we would make the nearest approaches possible to its possession, it must be by divorcing ourselves as much as in us lies from the flesh and its nature." In their spiritual views and apprehension of the nature of man, how these old heathens shame us!

The Scriptures teach us that God chose to reveal himself to his people chiefly in dreams, and we are entitled to conclude that the reason of this was, that the spirit was then more free to the reception of spiritual influences and impressions; and the class of dreams to which I next proceed, seem to be best explained by this hypothesis. It is also to be remarked, that the awe or fear which pervades a mortal at the mere conception of being brought into relation with a spirit, has no place in sleep, whether natural or magnetic. There is no fear then, no surprise; we seem to meet on an equality—is it not that we meet spirit to spirit? Is it not that our spirit being then released from the trammels—the dark chamber of the flesh, it does enjoy a temporary equality? Is not that true, that some German psychologist has said, "*The magnetic man is a spirit!*"

There are numerous instances to be met with, of persons receiving information in their sleep, which either is, or seems to be, communicated by their departed friends. The approach of danger, the period of the sleeper's death, or of that of some persons beloved, has been frequently made known in this form of dream.

Dr. Binns quotes, from Cardanus, the case of Johannes Maria Manrosenus, a Venetian senator, who, whilst Governor of Dalmatia, saw in a dream one of his brothers, to whom he was much attached; the brother embraced him and bade him farewell, because he was going into the other world; Maurosenus having followed him a long way weeping, awoke in tears and expressed much anxiety respecting this brother. Shortly afterwards he received tidings from Venice, that this Domatus, of whom he had dreamt, had died on the night and at the hour of the dream, of a pestilental fever, which had carried him off in three days.

On the night of the 21st of June, in the year 1813, a lady residing in the north of England, dreamt that her brother, who was then with his regiment in Spain, appeared to her saying, "Mary, I die this day at Vittoria."

Vittoria was a town which, previous to the famous battle, was not generally known even by name, in this country, and this dreamer, amongst others, had never heard of it; but, on rising, she eagerly resorted to a Gazetteer for the purpose of ascertaining if such a place existed. On finding that it was so, she immediately ordered her horses, and drove to the house of a sister, who resided some eight or nine miles off, and her first words on entering the room were, "Have you heard anything of John?" "No," replied the second sister, "but I know he is dead! He appeared to me last night, in a dream, and told me that he was killed at Vittoria. I have been looking into the Gazetteer and the Atlas, and I find there is such a place, and I am sure that he is dead!" And so it proved; the young man died that day at Vittoria, and, I believe, on the field of battle. If so, it is worthy of observation, that the communication was not made till the sisters slept.

A similar case to this, is that of Miss D., of G., who, one night, dreamt that she was walking about the washing greens, when a figure approached, which she recognized as that of a beloved brother, who was at that time with the British army, in America. It gradually faded away into a kind of anatomy, holding up its hands, through which the light could be perceived, and asking for clothes to dress a body for the grave. The dream recurred more than once in the same night, and, apprehending some misfortune, Miss D. noted down the date of the occurrence. In due course of post, the news arrived that this brother had been killed at the battle of Bunker's Hill. Miss D., who died only within the last few years, though unwilling to speak of the circumstance, never refused to testify to it as a fact.

Here, supposing this to be a real apparition, we see an instance of that desire for decent obsequies so constantly attributed by the ancients to the souls of the dead.

When the German poet, Collin, died at Vienna, a person named Hartmann, who was his friend, found himself very much distressed by the loss of a hundred and twenty florins, which he had paid for the poet, under a promise of reimbursement. As this sum formed a large portion of his whole possessions, the circumstance was occasioning him considerable anxiety, when he dreamt, one night, that his deceased friend appeared to him, and bade him immediately set two florins on No. 11, on the first calling of the little lottery, or loto, then about to be drawn. He was bade to confine his venture to two florins, neither less nor more; and to communicate this information to nobody. Hartmann availed himself of the hint, and obtained a prize of a hundred and thirty florins.

Since we look upon lotteries, in this country, as an immoral species of gambling, it may be raised as an objection to this dream, that such intelligence was an unworthy mission for a spirit, supposing the communication to have been actually made by Collin. But, in the first place, we have only to do with facts, and not with their propriety, or impropriety, according to our notions; and, by and by, I shall endeavour to show, that such discrepancies possibly arise from the very erroneous notions commonly entertained of the state of those who have disappeared from the terrestrial life.

Simonides, the poet, arriving at the sea-shore with the intention of embarking on board a vessel on the ensuing day, found an unburied body, which he immediately desired should be decently interred. On the same night, this deceased person appeared to him and bade him by no means go to sea, as he had proposed. Simonides obeyed the injunction, and beheld the vessel founder, as he stood on the shore. He raised a monument on the spot to the memory of his preserver, which is said still to exist, on which are engraven some lines to the effect, that it was dedicated by Simonides, the poet of Cheos, in gratitude to the dead who had preserved him from death.

A much esteemed secretary died a few years since, in the house of Mr. R. von N. About eight weeks afterwards, Mr. R. himself being ill, his daughter dreamt that the house-bell rang; and that on looking out, she perceived the secretary at the door. Having admitted him and enquired what he was come for, he answered, "to fetch somebody." Upon which, alarmed for her father, she exclaimed, "I hope not my father." He shook his head solemnly, in a manner that implied it was not the old man he had come for, and turned away towards a guest chamber, at that time vacant, and there disappeared at the door. The father recovered, and the lady left home for a few days, on a visit; on her return, she found her brother had arrived in the interval to pay a visit to his parents, and was lying sick in that room, where he died.

I will here mention a curious circumstance, regarding Mr. H., the gentleman alluded to in a former page, who, being at the sea-side, saw, in a dream, the danger that awaited his son when he went to bathe. This gentleman has frequently, on waking, felt a consciousness that he had been conversing with certain persons of his acquaintance—and, indeed, with some of whom he knew little—and has afterwards, not without a feeling of awe, learnt that these persons had died during the hours of his sleep.

Do not such circumstances entitle us to entertain the idea that I have above suggested, namely, that in sleep the spirit is free to see and to know, and to communicate with spirit, although the memory of this knowledge is rarely carried into the waking state.

The story of the two Arcadians, who travelled together to Megara, though reprinted in other works, I cannot omit here. One of these established himself, on the night of their arrival, at the house of a friend, whilst the other sought shelter in a public lodging-house for strangers. During the night, the latter appeared to the former, in a dream, and besought him to come to his assistance, as his villainous host was about to take his life, and only the most speedy aid could save him. The dreamer started from his sleep, and his first movement was to obey the summons, but, reflecting that it was only a dream, he presently lay down, and composed himself again to rest. But now his friend appeared before him a second time, disfigured by blood and wounds, conjuring him, since he had not listened to his first entreaties, that he would, at least, avenge his death. His host, he said, had murdered him, and was, at that moment, depositing his body in a dung-cart, for the purpose of conveying it out of the town. The dreamer was thoroughly alarmed, arose, and hastened to the gates of the city, where he found, waiting to pass out, exactly such a vehicle as his friend had described. A search being instituted, the body was found underneath the manure; and the host was, consequently, seized, and delivered over to the chastisement of the law.

"Who shall venture to assert," says Dr. Ennemoser, "that this communing with the dead in sleep is merely a subjective phenomenon, and that the presence of these apparitions is a pure illusion?"

A circumstance fully as remarkable as any recorded, occurred at Odessa, in the year 1842. An old blind man, named Michel, had for many years, been accustomed to get his living by seating himself every morning, on a beam, in one of the timber yards, with a wooden bowl at his feet, into which the passengers cast their alms. This long continued practice had made him well known to the inhabitants, and as he was believed to have been formerly a soldier, his blindness was attributed to the numerous wounds he had received in battle. For his own part, he spoke little, and never contradicted this opinion.

One night, Michel, by some accident, fell in with a little girl, of ten years old, named Powleska, who was friendless, and on the verge of perishing with cold and hunger. The old man took her home, and adopted her; and, from that time, instead of sitting in the timber yards, he went about the streets in her company, asking alms at the doors of the houses. The child called him *father*, and they were extremely happy together. But when they had pursued this mode of life for about five years, a misfortune befell them. A theft having been committed in a house which they had visited in the morning, Powleska was suspected and arrested, and the blind man was left once more alone. But, instead of resuming his former habits, he now disappeared altogether, and this circumstance causing the suspicion to extend to him, the girl was brought before the magistrate to be interrogated with regard to his probable place of concealment.

"Do you know where Michel is?" enquired the magistrate.

"He is dead!" replied she, shedding a torrent of tears.

As the girl had been shut up for three days, without any means of obtaining information from without, this answer, together with her unfeigned distress, naturally excited considerable surprise.

"Who told you he was dead?" they enquired.

"Nobody!"

"Then how can you know it?"

"I saw him killed!"

"But you have not been out of the prison?"

"But I saw it, nevertheless!"

"But how was that possible? Explain what you mean!"

"I cannot. All I can say is, that I saw him killed."

"When was he killed, and how?"

"It was the night I was arrested."

"That cannot be; he was alive when you were seized!"

"Yes, he was; he was killed an hour after that. They stabbed him with a knife."

"Where were you then?"

"I can't tell; but I saw it."

The confidence with which the girl asserted what seemed to her hearers impossible and absurd, disposed them to imagine that she was either really insane, or pretending to be so; so leaving Michel aside, they proceeded to interrogate her about the robbery, asking her if she was guilty.

"Oh, no!" she answered.

"Then how came the property to be found about you?"

"I don't know: I saw nothing but the murder."

"But there are no grounds for supposing Michel is dead; his body has not been found."

"It is in the aqueduct."

"And do you know who slew him?"

"Yes; it is a woman. Michel was walking very slowly, after I was taken from him A woman came behind him with a large kitchen-knife; but he heard her, and turned round; and then the woman flung a piece of grey stuff over his head, and struck him repeatedly with the knife, the grey stuff was much stained with the blood. Michel fell at the eighth blow, and the woman dragged the body to the aqueduct and let it fall in without ever lifting the stuff which stuck to his face."

As it was easy to verify these latter assertions, they dispatched people to the spot; and there the body was found with the piece of stuff over his head, exactly as she had described. But when they asked her how she knew all this, she could only answer "I don't know."

"But you know who killed him?"

"Not exactly: it is the same woman that put out his eyes; but, perhaps, he will tell me her name to-night; and if he does, I will tell it to you."

"Who do you mean by *he*?"

"Why, Michel, to be sure!"

During the whole of the following night, without allowing her to suspect their intention, they watched her; and it was observed that she never lay down, but sat upon the bed in a sort of lethargic slumber. Her body was quite motionless, except at intervals, when this repose was interrupted by violent nervous shocks, which pervaded her whole frame. On the ensuing day, the moment she was brought before the judge, she declared that she was now able to tell them the name of the assassin.

"But stay, said the magistrate; "did Michel never tell you, when he was alive, how he lost his sight?"

"No; but the morning before I was arrested, he promised me to do so; and that was the cause of his death."

"How could that be?"

"Last night Michel came to me, and he pointed to the man hidden behind the scaffolding on which he and I had been sitting. He showed me the man listening to us, when he said, 'I'll tell you all about that to-night;' and then the man——"

"Do you know the name of this man?"

"It is *Luck;* he went afterwards to a broad

street that leads down to the harbour, and he entered the third house on the right——"

"What is the name of the street?"

"I don't know: but the house is one story lower than the adjoining ones. Luck told Catherine what he had heard, and she proposed to him to assassinate Michel; but he refused, saying, 'it was bad enough to have burnt out his eyes fifteen years before, whilst he was asleep at your door, and to have kidnapped him into the country.' Then I went in to ask charity, and Catherine put a piece of plate into my pocket, that I might be arrested: then she hid herself behind the aqueduct to wait for Michel, and she killed him."

"But, since you say all this, why did you keep the plate?—why didn't you give information?"

"But I didn't see it then. Michel showed it me last night."

"But what should induce Catherine to do this?"

"Michel was her husband, and she had forsaken him to come to Odessa and marry again. One night, fifteen years ago, she saw Michel, who had come to seek her. She slipped hastily into her house, and Michel, who thought she had not seen him, lay down at her door to watch; but he fell asleep, and then Luck burnt out his eyes, and carried him to a distance."

"And is it Michel who has told you this?"

"Yes: he came, very pale and covered with blood; and he took me by the hand and showed me all this with his fingers."

Upon this, Luck and Catherine were arrested; and it was ascertained that she had actually been married to Michel in the year 1819, at Kherson. They at first denied the accusation, but Powleska insisted, and they subsequently confessed the crime. When they communicated the circumstances of the confession to Powleska, she said, "I was told it last night."

This affair naturally excited great interest, and people all round the neighbourhood hastened into the city to learn the sentence.

CHAPTER VI.

DOUBLE DREAMING AND TRANCE.

AMONGST the phenomena of the dream-life which we have to consider, that of double-dreaming forms a very curious department. A somewhat natural introduction to this subject may be found in the cases above recorded of Professor Herder and Mr. S. of Edinburgh, who appear in their sleep to have received so lively an impression of those earnest wishes of their dying friends to see them, that they found themselves irresistably impelled to obey the spiritual summons. These two cases occurred to men engaged in active daily life, and in normal physical conditions, on which account I particularly refer to them here, although many similar ones might be adduced.

With respect to this subject of double-dreaming, Dr. Ennemoser thinks that it is not so difficult to explain as might appear on a first view, since he considers that there exists an indisputable sympathy betwixt certain organisms, especially where connected by relationship, or by affection, which may be sufficient to account for the supervention of simultaneous thoughts, dreams, or presentiments; and I have met with some cases where the magnetiser and his patient have been the subjects of this phenomenon. With respect to the power asserted to have been frequently exercised by causing or suggesting dreams by an operator at a distance from the sleeper, Dr. E. considers the two parties to stand in a positive and negative relation to each other; the antagonistic power of the sleeper being = 0, he becomes a perfectly passive recipient of the influence exerted by his positive *half*, if I may use the expression; for, where such a polarity is established, the two beings seem to be almost blended into one; whilst Dr. Passavent observes, that we cannot pronounce what may be the limits of the nervous force, which certainly is not bounded by the termination of its material conductors.

I have yet myself met with no instance of dream compelling by a person at a distance; but Dr. Ennemoser, says, that Agrippa von Nettesheim asserts that this can assuredly be done, and also that the Abbot Trithemius, and others, possessed the power. In modern times, Wesermaun, in Dusseldorf, pretended to the same faculty, and affirms that he had frequently exercised it.

All such phenomena, Dr. Passavent attributes to the interaction of imponderables—or of one universal imponderable under different manifestations—which acts not only within the organism, but beyond it, independently of all material obstacles; just as a sympathy appears betwixt one organ and another, unobstructed by the intervening ones; and he instances the sympathy which exists betwixt the mother and the fœtus, as an example of this sort of double life, and standing as midway betwixt the sympathy between two organs in the same body and that between two separate bodies; each having its own life, and its life also in and for another, as parts of one whole. The sympathy betwixt a bird and the eggs it sits upon is of the same kind; many instances having been observed, wherein eggs taken from one bird and placed under another, have produced a brood feathered like the foster, instead of the real parent.

Thus, this vital force may extend dynamically the circle of its influence, till, under favourable circumstances, it may act on other organisms making their organs its own.

I need scarcely remind my readers of the extraordinary sympathies manifested by the Siamese twins—Chang and Eng. I never saw them myself; and, for the benefit of others in the same situation, I quote the following particulars from Dr. Passavent:—"They were united by a membrane which extended from the breast-bone to the navel; but, in other respects, were not different from their countrymen in general. They were exceedingly alike, only that Eng was rather the most robust of the two. Their pulsations were not always coincident. They were active and agile, and fond of bodily exercises; their intellects were well-developed, and their tones of voice and accent were precisely the same. As they never conversed together, they had nearly forgotten their native tongue. If one was addressed, they both answered. They played some games of skill, but never with each other; as that, they said, would have been like the right hand playing with the left. They read the same book at the same time, and sang together in unison. In America they had a fever, which ran precisely a similar course with each. Their hunger, thirst, sleeping and waking, were always coincident; and their tastes and inclinations were identical. Their movements were so simultaneous that it was impossible to distinguish with which the impulse had originated; they appeared to have but one will. The idea of being separated by an operation was abhorrent to them; and they consider themselves much happier in their duality than are the individuals who look upon them with pity.

This admirable sympathy, although necessarily in an inferior degree, is generally manifested, more or less, betwixt all persons twin born. Dr. Passavent, and other authorities, mention several instances of this kind, in which, although at some distance from each other, the same malady appeared simultaneously in both, and ran precisely a similar course. A very affecting instance of this sort of sympathy was exhibited, not very long ago, by a young lady, twin-born, who was suddenly seized with an unaccountable horror, followed by a strange convulsion, which the doctor, who was hastily called in, said, exactly resembled the struggles and sufferings of a person drowning. In process of time, the news arrived, that her twin brother, then abroad, had been drowned precisely at that period.

It is, probably, a link of the same kind, that is established betwixt the magnetiser and his patient, of which, besides those recorded in various works on the subject, some curious instances have come to my knowledge, such as uncontrollable impulses to go to sleep, or to perform certain actions, in subserviance to the will of the distant operator. Mr. W. W., a gentleman well known in the north of England, related to me, that he had been cured, by magnetism, of a very distressing malady, During part of the process of cure, after the *rapport* had been well established, the operations were carried on whilst he was at Malvern, and his magnetiser at Cheltenham, under which circumstances the existence of this extraordinary dependence was frequently exhibited in a manner that left no possibility of doubt. On one occasion, I remember, that Mr. W. W. being in the magnetic sleep, he suddenly started from his seat, clasping his hands as if startled, and, presently afterwards, burst into a violent fit of laughter. As, on waking, he could give no account of these impulses, his family wrote to the magnetiser to enquire if he had sought to excite any particular manifestations in his patient, as the sleep had been somewhat disturbed. The answer was, that no such intention had been entertained, but that the disturbance might possibly have arisen from one to which he had himself been subjected. "Whilst my mind was concentrated on you," said he, "I was suddenly so much startled by a violent knock at the door, that I actually jumped off my seat, clasping my hands with affright. I had a hearty laugh at my own folly, but am sorry if you were made uncomfortable by it."

I have met with some accounts of a sympathy of this kind existing betwixt young children and their parents, so that the former have exhibited great distress and terror at the moment that death or danger have supervened to the latter; but it would require a great number of instances to establish this particular fact, and separate it from cases of accidental coincidence. Dr. Passavent, however, admits the phenomena.

I shall return to these mysterious influences by and by; but to revert, in the meanwhile, to the subject of double dreams, I will relate one that occurred to two ladies, a mother and daughter, the latter of whom related it to me. They were sleeping in the same bed at Cheltenham, when the mother, Mrs. C., dreamt that her brother-in-law, then in Ireland, had sent for her; that she entered his room, and saw him in bed, apparently dying. He requested her to kiss him, but, owing to his livid appearance, she shrank from doing so, and awoke with the horror of the scene upon her. The daughter awoke at the same moment, saying, "Oh, I have had such a frightful dream!" "Oh, so have I!" returned the mother; "I have been dreaming of my brother-in-law!" "My dream was about him, too," added Miss C. "I thought I was sitting in the drawing-room, and that he came in wearing a shroud, trimmed with black ribbons, and, approaching me, he said, 'My dear niece, your mother has refused to kiss me, but I am sure you will not be so unkind!'"

As these ladies were not in habits of regular correspondence with their relative, they knew that the earliest intelligence likely to reach them, if he were actually dead, would be by means of the Irish papers; and they waited anxiously for the following Wednesday, which was the day these journals were received in Cheltenham. When that morning arrived, Miss C. hastened at an early hour to the reading-room, and there she learnt what the dreams had led them to expect: their friend was dead; and they afterwards ascertained that his decease had taken place on that night. They moreover observed, that neither one or the other of them had been speaking or thinking of this gentleman for some time previous to the occurrence of the dreams; nor had they any reason whatever for uneasiness with regard to him. It is a remarkable peculiarity in this case, that the dream of the daughter appears to be a continuation of that of the mother. In the one, he is seen alive; in the other, the shroud and black ribbons seem to indicate that he is dead; and he complains of the refusal to give him a farewell kiss.

One is almost inevitably led here to the conclusion that the thoughts and wishes of the dying man were influencing the sleepers; or, that the released spirit was hovering near them.

Pomponius Mela relates, that a certain people in the interior of Africa lay themselves down to sleep on the graves of their forefathers, and believe the dreams that ensue to be the unerring counsel of the dead.

The following dream, from St. Austin, is quoted by Dr. Binns :—" Præstantius desired, from a certain philosopher, the solution of a doubt, which the latter refused to give him; but in the following night, the philosopher appeared at his bed-side and told him what he desired to know. On being asked, the next day, why he had chosen that hour for his visit, he answered, 'I came not to you truly, but in my dream I appeared to you to do so.' In this case, however, only one of the parties seems to have been asleep; for Præstantius says that he was awake; and it is, perhaps, rather an example of another kind of phenomena, similar to the instance recorded of himself by the late Joseph Wilkins, a dissenting minister; who says, that being one night asleep, he dreamed that he was travelling to London, and that as it would not be much out of his way, he would go by Gloucestershire and call upon his friends. Accordingly he arrived at his father's house, but finding the front door closed, he went round to the back, and there entered. The family, however, being already in bed, he ascended the stairs, and entered his father's bedchamber. Him he found asleep; but to his mother, who was awake, he said, as he walked round to her side of the bed, 'Mother,

I am going a long journey, and am come to bid you good bye;' to which she answered, 'Oh, dear son, thee art dead!' Though struck with the distinctness of the dream, Mr. Wilkins attached no importance to it, till, to his surprise, a letter arrived from his father, addressed to himself, if alive; or, if not, to his surviving friends, begging earnestly for immediate intelligence; since they were under great apprehensions that their son was either dead or in danger of death; for that on such a night (naming that on which the above dream had occurred), he, the father being asleep and Mrs. W. awake, she had distinctly heard somebody try to open the fore door, which being fast, the person had gone round to the back and there entered. She had perfectly recognised the footstep to be that of her son, who had ascended the stairs, and entering the bedchamber had said to her, 'Mother, I am going a long journey, and am come to bid you good bye;' whereupon, she had answered, 'Oh, dear son, thee art dead!' Much alarmed, she had awakened her husband and related what had occurred, assuring him that it was not a dream, for that she had not been asleep at all. Mr. W. mentions that this curious circumstance took place in the year 1754, when he was living at

Ottery; and that he had frequently discussed the subject with his mother, on whom the impression made was even stronger than on himself. Neither death, nor anything else remarkable ensued."

A somewhat similar instance to this, which I also quote from Dr. Binns, is that of a gentleman who dreamt that he was pushing violently against the door of a certain room in a house with which he was well acquainted, whilst the people in that room were, at the same time, actually alarmed by a violent pushing against the door, which it required their utmost force effectually to resist. As soon as the attempt to burst open the door had ceased, the house was searched; but nothing discovered to account for the disturbance.

These examples are extremely curious; and they conduct us by a natural transition to another department of this mysterious subject.

There must be few persons who have not heard amongst their friends and acquaintance instances of what is called a *Wraith*—that is, that in the moment of death, a person is seen in a place where *bodily* he is not. I believe the Scotch use this term also in the same sense as the Irish word *Fetch;* which is a person's double seen at some indefinite period previous to his death, of which such an appearance is generally supposed to be a prognostic. The Germans express the same thing by the word Döppelganger.

With respect to the appearance of wraiths, at the moment of death, the instances to be met with are so numerous and well authenticated, that I generally find the most sceptical people unable to deny that some such phenomenon exists, although they evade, without, I think, diminishing the difficulty, by pronouncing it to be of a subjective, and not of an objective, nature; that is, that the image of the dying person is, by some unknown operation, presented to the imagination of the seer, without the existence of any real outstanding figure, from which it is reflected; which reduces such instances so nearly to the class of mere sensuous illusion, that it seems difficult to draw the distinction. The distinction these theorists wish to imply, however, is, that the latter are purely subjective and self-originating, whilst the others have an external cause, although not an external visible object—the image seen being protruded by the imagination of the seer, in consequence of an unconscious intuition of the death of the person whose wraith is perceived.

Instances of this kind of phenomenon have been common in all ages of the world, insomuch that Lucretius, who did not believe in the immortality of the soul, and was yet unable to deny the facts, suggested the strange theory that the superficial surfaces of all bodies were continually flying off, like the coats of an onion, which accounted for the appearance of wraiths, ghosts, doubles, &c.; and a more modern author, Gaffarillus, suggests that corrupting bodies send forth vapours, which, being compressed by the cold night air, appear visible to the eye in the forms of men.

It will not be out of place, here, to mention the circumstance recorded in Professor Gregory's Abstract of Baron Von Reichenbach's Researches in Magnetism, regarding a person called Billing, who acted in the capacity of amanuensis to the blind poet Pfeffel, at Colmar. Having treated of various experiments, by which it was ascertained that certain sensitive persons were not only able to detect electric influences of which others were unconscious, but could also perceive, emanating from the wires and magnets, flames which were invisible to people in general; "the Baron," according to Dr. Gregory, "proceeded to a useful application of the results, which is, says he, so much the more welcome, as it utterly eradicates one of the chief foundations of superstition, that worst enemy to the development of human enlightenment and liberty. A singular occurrence, which took place at Col mar, in the garden of the poet Pfeffel, has been made generally known by various writings. The following are the essential facts. The poet, being blind, had employed a young clergyman, of the evangelical church, as amanuensis. Pfeffel, when he walked out, was supported and led by this young man, whose name was Billing. As they walked in the garden, at some distance from the town, Pfeffel observed, that, as often as they passed over a particular spot, the arm of Billing trembled, and he betrayed uneasiness. On being questioned, the young man reluctantly confessed that, as often as he passed over that spot, certain feelings attacked him, which he could not controul, and which he knew well, as he always experinced the same, in passing over any place where human bodies lay buried. He added, that, at night, when he came near such places, he saw supernatural appearances. Pfeffel, with the view of curing the youth of what he looked on as a fancy, went that night with him to the garden. As they approached the spot in the dark, Billing perceived a feeble light, and when still nearer, he saw a luminous ghost-like figure floating over the spot. This he described as a female form, with one arm laid across the body, the other hanging down, floating in the upright posture, but tranquil, the feet only a hand-breadth or two above the soil. Pfeffel went alone, as the young man declined to follow him, up to the place where the figure was said to be, and struck about in all directions with his stick, besides running actually through the shadow; but the figure was not more affected than a flame would have been: the luminovs form, according to Billing, always returned to its original position after these experiments. Many things were tried during several months, and numerous companies of people were brought to the spot, but the matter remained the same, and the ghost seer adhered to his serious assertion, and to the opinion founded on it, that some individual lay buried there. At last, Pfeffel had the place dug up. At a considerable depth was found a firm layer of white lime, of the length and breadth of a grave, and of considerable thickness, and when this had been broken into, there were found the bones of a human being. It was evident that some one had been buried in the place, and covered with a thick layer of lime (quicklime), as is generally done in times of pestilence, of earthquakes, and other similar events. The bones were removed, the pit filled up, the lime mixed and scattered abroad, and the surface again made smooth. When Billing was now brought back to the place, the phenomena did not return, and the nocturnal spirit had for ever disappeared.

"It is hardly necessary to point out to the reader what view the author takes of this story, which excited much attention in Germany, because it came from the most truthful man alive, and theologians and psychologists gave to it sundry terrific meanings. It obviously falls into the province of chemical action, and thus meets with a simple and clear explanation from natural and physical causes. A corpse is a field for abundant chemical changes, decompositions, fermentation, putrefaction, gasification, and general play of affinities. A stratum of quicklime, in a narrow pit, unites its powerful affinities to those of the organic matters, and gives rise to a long continued working of the whole. Rain-water filters through and contributes to the action: the lime on the outside of the mass first falls to a fine powder, and afterwards, with more water, forms lumps which are very slowly penetrated by the air. Slaked lime prepared for building, but not used, on account of some cause connected with a warlike state of society some centuries since, has been found in subterraneous holes or pits, in the ruins of old castles; and the mass, except on the outside, was so unaltered, that it has been used for modern buildings. It is evident, therefore, that in such circumstances there must be a very slow and long continued chemical action, partly owing to the slow penetration of the mass of lime by the external carbonic acid, partly to the changes going on in the remains of animal matter, at all events as long as any is left. In the above case, this must have gone on in Pfeffel's garden, and, as we know that chemical action is invariably associated with light, visible to the sensitive, this must have been the origin of the luminous appearance, which again must have continued until the mutual affinities of the organic remains, the lime, the air, and water, had finally come to a state of chemical rest or equilibrium. As soon, therefore, as a sensitive person, although otherwise quite healthy, came that way, and entered within the sphere of the force in action, he must feel, by day, like Mdlle. Maix, the sensations so often described, and see by night, like Mdlle. Reichel, the luminous appearance. Ignorance, fear, and superstition, would now dress up the feebly shining vapourous light into a human form, and furnish it with human limbs and members; just as we can at pleasure fancy every cloud in the sky to represent a man or a demon.

"The wish to strike a fatal blow at the monster superstition, which, at no distant period, poured out on European society from a similar source, such inexpressible misery, when, in trials for witchcraft, not hundreds, not thousands, but hundreds of thousands of innocent human beings perished miserably, either on the scaffold, at the stake, or by the effects of torture,—this desire induced the author to try the experiment of bringing, if possible, a highly sensitive patient, by night, to a churchyard. It appeared possible that such a person might see, over graves, in which mouldering bodies lie, something similar to that which Billing had seen. Mdlle. Reichel had the courage, rare in her sex, to gratify this wish of the author. On two very dark nights she allowed herself to be taken from the castle of Reisenberg, where she was living, with the author's family, to the neighbouring churchyard of Grunzing. The result justified his anticipation in the most beautiful manner.

She very soon saw a light, and observed on one of the graves, along its length, a delicate, breathing flame: she also saw the same thing, only weaker, on a second grave. But she saw neither witches nor ghosts; she described the fiery appearance as a shining vapour, one to two spans high, extending as far as the grave, and floating near its surface. Some time afterwards she was taken to two large cemeteries near Vienna, where several burials occur daily, and graves lie about by thousands. Here she saw numerous graves provided with similar lights. Wherever she looked, she saw luminous masses scattered about. But this appearance was most vivid over the newest graves, while in the oldest it could not be perceived. She described the appearance less as a clear flame, than as a dense vaporous mass of fire, intermediate between fog and flame. On many graves the flame was four feet high, so that when she stood on them, it surrounded her up to the neck. If she thrust her hand into it, it was like putting it into a dense fiery cloud. She betrayed no uneasiness, because she had all her life been accustomed to such emanations, and had seen the same, in the author's experiments, often produced by natural causes. Many ghost stories will now find their natural explanation. We can also see, that it was not altogether erroneous, when old women declared that all had not the gift to see the departed wandering about their graves; for it must have always been the sensitive alone who were able to perceive the light given out by the chemical action going on in the corpse. The author has thus, he hopes, succeeded in tearing down one of the most impenetrable barriers erected by dark ignorance and superstitious folly, against the progress of natural truth."

"[The reader will at once apply the above most remarkable experiments to the explanation of corpse-lights in church-yards, which were often visible to the gifted alone, to those who had the second sight, for example. Many nervous or hysterical females must often have been alarmed by white, faintly luminous objects, in dark churchyards, to which objects fear has given a defined form. In this, as well as in numerous other points, which will force themselves on the attention of the careful reader of both works, Baron Reichenbach's experiments illustrate the experiences of the Seeress of Prevorst.—W. G.]" *

* This very curious work I have translated from the German. Published by Moore, London.—C. C.

That the flames here described may have originated in chemical action, is an opinion I have no intention of disputing; the fact may possibly be so; such a phenomenon has frequently been observed hovering over coffins and decomposing flesh; but I confess I cannot perceive the slightest grounds for the assertion that it was the ignorance, fear, and superstition of Billing, who was an Evangelical clergyman, that caused him to dress up this vaporous light in a human form and supply it with members, &c. In the first place, I see no proof adduced that Billing was either ignorant or superstitious, nor even afraid; the feelings he complained of, appearing to be rather physical than moral; and it must be a weak person indeed, who, in company with another, could be excited to such a freak of the imagination. It is easily comprehensible, that that which appeared only a luminous vapour by day, might when reflected on a darker atmosphere, present a defined form; and the suggestion of this possibility might lead to some curious speculations, with regard to a mystery called the palinganesia, said to have been practised by some of the chemists and alchemists of the sixteenth century.

Gaffarillus, in his book, entitled "*Curiosites*

Inouies," published in 1650, when speaking on the subject of talismans, signatures, &c., observes, that since in many instances the plants used for these purposes were reduced to ashes, and no longer retained their form, their efficacy which depended on their figure should inevitably be destroyed; but this, he says, is not the case, since, by an admirable potency existing in nature, the form, though invisible, is still retained in the ashes. This, he observes, may appear strange to those who have never attended to the subject; but he asserts that an account of the experiment will be found in the works of Mr. Du Chesne. one of the best chemists of the period, who had been shown, by a Polish physician, at Cracow, certain phials containing ashes, which, when duly heated, exhibited the forms of various plants. A small obscure cloud was first observed, which gradually took on a defined form, and presented to the eye a rose, or whatever plant or flower the ashes consisted of. Mr. Du Chesne, however, had never been able to repeat the experiment, though he had made several unsuccessful attempts to do so; but at length he succeeded, by accident, in the following manner:—Having for some purpose extracted the salts from some burnt nettles, and having left the lie outside the house, all night, to cool, in the morning he found it frozen; and, to his surprise, the form and figure of the nettles were so exactly represented on the ice, that the living plant could not be more perfect. Delighted at this discovery, he summoned Mr. De Luynes, parliamentary councillor, to behold this curiosity; from whence, he says, they both concluded, that when a body dies, its form or figure still resides in its ashes.

Kircher, Vallemont, Digby, and others, are said to have practised this art of resuscitating the forms of plants from their ashes; and at the meeting of naturalists at Stuttgard, in 1834, a Swiss savant seems to have revived the subject, and given a receipt for the experiment extracted from a work by Oetinger, called "Thoughts on the Birth and Generation of Things." "The earthly husk," says Oetinger, "remains in the retort, whilst the volatile essence ascends, like a spirit, perfect in form, but void of substance."

But Oetinger also records another discovery of this description, which, he says, he fell upon unawares. A woman having brought him a large bunch of balm, he laid it under the tiles, which were yet warm with the summer's heat, where it dried in the shade. But, it being in the month of September, the cold soon came, and contracted the leaves, without expelling the volatile salts. They lay there till the following June, when he chopped up the balm, put it into a glass retort, poured rain water upon it, and placed a receiver above. He afterwards heated it till the water boiled, and then increased the heat; whereupon there appeared, on the water, a coat of yellow oil, about the thickness of the back of a knife, and this oil shaped itself into the forms of innumerable balm leaves, which did not run one into another, but remained perfectly distinct and defined, and exhibited all the marks that are seen in the leaves of the plant. Oetinger says he kept the fluid some time, and showed it to a number of people. At length, wishing to throw it away, he shook it, and the leaves ran into one another with the disturbance of the oil, but resumed their distinct shape again, as soon as it was at rest, the fluid form retaining the perfect signature.

Now, how far these experiments are really practicable, I connot say, their not being repeated, or not being repeated successfully, is no very decided argument against their possibility, as all persons acquainted with the annals of chemistry well know; but there is, certainly, a curious coincidence betwixt these details, and the experience of Billing; where it is to be observed, that, according to his account—and what right have we to dispute it—the figure after being disturbed by Pfeffel, always resumed its original form. The same peculiarity has been observed with respect to some apparitions, where the spectator has been bold enough to try the experiment. In a letter to Dr. Bentley, from the Rev. Thos. Wilkins, curate of Warblington, in Hampshire, written in the year 1695, wherein he gives an account of an apparition which haunted the parsonage house, and which he himself, and several other persons, had seen; he particularly mentions that, thinking it might be some fellow hid in the room, he put his arm out to feel it, and his hand seemingly went through the body of it, and felt no manner of substance. until it reached the wall, "then I drew back my hand, but still the apparition was in the same place."

Yet this spectre did not appear above, or near a grave, but moved from place to place, and gave considerable annoyance to the inhabitants of the rectory.

With respect to the lights over the graves, sufficing to account for the persuasion regarding what is called corpse candles, they certainly, up to a certain point, afford a very satisfactory explanation, but that explanation does not comprehend the whole of the mystery, for most of those persons who have professed to see corpse candles, have also asserted that they were not always stationery over the graves, but sometimes moved from place to place, as in the following instance, which was related to me by a gentleman who assured me he received the account from the person who witnessed the phenomenon. Now, this last fact, I mean the locomotion of the lights, will, of course, be disputed; but so was their existence; yet they exist, for all that, and may travel from place to place, for anything we know to the contrary.

The story related to me, or a similar instance, is, I think, mentioned by Mrs. Grant; but it was to the effect that a minister, newly inducted in his cure, was standing one evening leaning over the wall of the church-yard which adjoined the manse, when he observed a light hovering over a particular spot. Supposing it to be somebody with a lanthorn, he opened the wicket, and went forward to ascertain who it might be; but before he reached the spot the light moved onwards; and he followed, but could see nobody. It did not rise far from the ground, but advanced rapidly across the road, entered a wood, and ascended a hill, till it at length disappeared at the door of a farm-house. Unable to comprehend of what nature this light could be, the minister was deliberating whether to make enquiries at the house or return, when it appeared again, seeming to come out of the house, accompanied by another, passed him, and going over the same ground, they both disappeared on the spot where he had first observed the phenomenon. He left a mark on the grave by which he might recognise it, and the next day enquired of the sexton whose it was. The man said, it belonged to a family that lived up the hill, indicating the house the light had stopped at, named M'D.—but that it was a considerable time since any one had been buried there. The minister was extremely surprised to learn, in the course of the day, that a child of that family had died of scarlet fever on the preceding evening. With respect to the class of phenomena accompanied by this phosphorescent light, I shall have more to say by and by. The above will appear a very incredible story to many people, and there was a time that it would have appeared equally so to myself;

but I have met with so much strange corroborative evidence, that I no longer feel myself entitled to reject it. I asked the gentleman who told me the story, whether he believed it; he said that he could not believe in anything of the sort. I then enquired if he would accept the testimony of that minister on any other question, and he answered, "Most assuredly." As, however, I shall have occasion to recur to this subject in a subsequent chapter, I will leave it aside for the present, and relate some of the facts which led me to the consideration of the above theories and experiments. Dr. S. relates, that a Madame T., in Prussia, dreamt, on the 16th March, 1832, that the door opened, and her godfather, Mr. D., who was much attached to her, entered the room, dressed as he usually was when prepared for church on Sundays; and that, knowing him to be in bad health, she asked him what he was doing abroad at such an early hour, and whether he was quite well again. Whereupon, he answered, that he was; and, being about to undertake a very long journey, he had come to bid her farewell, and to intrust her with a commission, which was, that she would deliver a letter he had written to his wife; but accompanying it with an injunction that she, the wife, was not to open it till that day four years, when he would return himself, precisely at five o'clock in the morning, to fetch the answer, till which period he charged her not to break the seal. He then handed her a letter, sealed with black, the writing on which shone through the paper, so that she, the dreamer, was able to perceive that it contained an announcement to Mrs. D., the wife, with whom, on account of the levity of her character, he had long lived unhappily, that she would die that time four years. At this moment, the sleeper was awakened by what appeared to her a pressure of the hand, and, feeling an entire conviction that this was something more than an ordinary dream, she was not surprised to learn that her godfather was dead. She related the dream to Madame D., omitting, however, to mention the announcement contained in the letter, which she thought the dream plainly indicated was not to be communicated. The widow laughed at the story, soon resumed her gay life, and married again. In the winter of 1835-6, however, she was attacked by an intermittent fever, on which occasion Dr. S. was summoned to attend her. After various vicissitudes, she finally sunk; and, on the 16th of March, 1836, exactly at five o'clock in the morning, she suddenly started up in her bed, and, fixing her eyes apparently on some one she saw standing at the foot, she exclaimed, "What are you come for? God be gracious to me! I never believed it!" She then sank back, closed her eyes, which she never opened again, and, in a quarter of an hour afterwards, expired very calmly.

A friend of mine, Mrs. M., a native of the West Indies, was at Blair Logie, at the period of the death of Dr. Abercrombie, in Edinburgh, with whom she was extremely intimate. Dr. A. died quite suddenly, without any previous indisposition, just as he was about to go out in his carriage, at eleven o'clock on a Thursday morning. On the night between the Thursday and Friday, Mrs. M. dreamt that she saw the family of Dr. A. all dressed in white, dancing a solemn funereal dance, upon which she awoke, wondering that she should have dreamt a thing so incongruous, since it was contrary to their custom to dance on any occasion. Immediately afterwards, whilst speaking to her maid, who had come to call her, she saw Dr. Abercrombie against the wall, with his jaw fallen, and a livid countenance, mournfully shaking his head, as he looked at her. She passed the day in great uneasiness, and wrote to enquire for the Doctor, relating what had happened, and expressing her certainty that he was dead; the letter was seen by several persons in Edinburgh, on the day of its arrival.

The two following cases seem rather to belong to what is called in the East *Second Hearing*, although sympathy was probably the exciting cause of the phenomena. A lady and gentleman, in Berwickshire, were awakened one night by a loud cry, which they both immediately recognised to proceed from the voice of their son, who was then absent, and at a considerable distance. Tidings subsequently reached them that exactly at that period their son had fallen overboard and was drowned; and on another occasion, in Perthshire, a person aroused her husband, one night, saying that their son was drowned, for she had been awakened by the splash. Her presentiment also proved too well founded, the young man having fallen from the mast-head of the ship. In both cases we may naturally conclude, that the thoughts of the young men, at the moment of the accident, would rush homewards; and, admitting Dr. Ennemoser's theory of polarity, the passive sleepers became the recipients of the force. I confess, however, that the opinions of another section of philosophers appear to me more germain to the matter; although to many persons they will doubtless be difficult of acceptance, from their appertaining to those views commonly called mystical.

These psychologists then believe, as did Socrates and Plato, and others of the ancients, that in certain conditions of the body, which conditions may arise naturally, or be produced artificially, the links which unite it with the spirit may be more or less loosened; and that the latter may thus be temporarily disjoined from the former, and so enjoy a foretaste of its future destiny. In the lowest, or first degree, of this disunion, we are awake, though scarcely conscious, whilst the imagination is vivified to an extraordinary amount, and our fancy supplies images almost as lively as the realities. This, probably, is the temporary condition of inspired poets and eminent discoverers.

Sleep is considered another stage of this disjunction, and the question has even been raised, whether, when the body is in profound sleep, the spirit is not altogether free and living in another world, whilst the organic life proceeds as usual, and sustains the temple till the return of its inhabitant. Without, at present, attempting to support or refute this doctrine, I will only observe, that once admitting the possibility of the disunion, all consideration of *time* must be set aside as irrelavent to the question; for spirit, freed from matter, must move with the rapidity of thought—in short, *a spirit must be where its thoughts and affections are.*

It is the opinion of these psychologists, however, that in the normal and healthy condition of man, the union of body, soul, and spirit, is most complete; and that all the degrees of disunion in the waking state are degrees of morbid derangement. Hence it is, that somnambulists and clairvoyantes are chiefly to be found amongst sickly women There have been persons who have appeared to possess a power which they could exert at will, whereby they withdrew from their bodies, these remaining during the absence of the spirit in a state of catalepsy scarcely, if at all, to be distinguished from death.

I say *withdrew from their bodies*, assuming that to be the explanation of the mystery; for, of course, it is but an assumption. Epimenides is recorded to have possessed this faculty, and Hermotinus, of Clazomenes, is said to have wandered, in spirit, over the world, whilst his body lay apparently dead. At length, his wife taking advantage of this absence of his soul, burnt his body, and thus intercepted its return. So say Lucien and Pliny, the elder; and Varro relates, that the eldest of two brothers, named Corfidius, being supposed to die, his will was opened and preparations were made for his funeral by the other brother, who was declared his heir. In the mean time, however, Corfidius revived, and told the astonished attendants, whom he summoned by clapping his hands, that he had just come from his younger brother, who had committed his daughter to his care, and informed him where he had buried some gold, requesting that the funeral preparations he had made might be converted to his own use. Immediately afterwards, the news arrived that the younger brother was unexpectedly deceased, and the gold was found at the place indicated The last appears to have been a case of natural trance; but the two most remarkable instances of voluntary trance I have met with in modern times is that of Colonel Townshend, and the Dervish, who allowed himself to be buried. With regard to the former, he could, to all appearance, die whenever he pleased; his heart ceased to beat; there was no perceptible respiration; and his whole frame became cold and rigid as death itself; the features being shrunk and colourless, and the eyes glazed and ghastly. He would continue in this state for several hours, and then gradually revive; but the revival does not appear to have been an effort of will; or rather, we are not informed whether it was so or not. Neither are we told whether he brought any recollections back with him, nor how this strange faculty was first developed or discovered—all very important points, and well worthy of investigation. He seems to have made this experiment, however, once too often; for, on one of these occasions, he was found to have actually expired.

With respect to the Dervish or Fakeer, an account of his singular faculty, was, I believe, first presented to the public in the Calcutta papers, about nine or ten years ago. He had then frequently exhibited it for the satisfaction of the natives, but subsequently he was put to the proof by some of the European officers and residents. Captain Wade, political agent, at Loodhiana, was present when he was disinterred, ten months after he had been buried by General Ventura, in presence of the Maharajah and many of his principal Sirdars.

It appears that the man previously prepared himself by some processes, which, he says, temporarily annihilate the powers of digestion, so that milk received into the stomach undergoes no change. He next forces all the breath in his body into his brain, which becomes very hot, upon which the lungs collapse, and the heart ceases to beat. He then stops up, with wax, every aperture of the body through which air could enter, except the mouth, but the tongue is so turned back as to close the gullet, upon which a state of insensibility ensues. He is then stripped and put into a linen bag, and, on the occasion in question, this bag was sealed with Runjeet Sing's own seal. It was then placed in a deal box, which was also locked and sealed, and the box being buried in a vault, the earth was thrown over it and trod down, after which a crop of barley was sown on the spot, and sentries placed to watch it. The Maharajah, however, was so sceptical, that, in spite of all these precautions, he had him, twice in the course of the ten months, dug up and examined; and each time he was found to be exactly in the same state as when they had shut him up.

When he is disinterred, the first step towards his recovery is to turn back his tongue, which is found quite stiff, and requires for some time to be retained in its proper position by the finger; warm water is poured upon him, and his eyes and lips moistened with ghee, or oil. His recovery is much more rapid than might be expected, and he is soon able to recognise the bystanders, and converse. He says, that, during this state of trance, his dreams are ravishing, and that it is very painful to be awakened, but I do not know that he has ever disclosed any of his experiences. His only apprehension seems to be, lest he should be attacked by insects, to avoid which accident the box is slung to the ceiling. The interval seems to be passed in a complete state of Hibernation; and when he is taken up, no pulse is perceptible, and his eyes are glazed like those of a corpse.

He subsequently refused to submit to the conditions proposed by some English officers, and thus incurred their suspicions, that the whole thing was an imposition; but the experiment has been too often repeated by people very well capable of judging, and under too stringent precautions, to allow of this mode of escaping the difficulty. The man assumes to be *holy*, and is very probably a worthless fellow, but that does not affect the question one way or the other. Indian princes do not permit themselves to be imposed on with impunity; and, as Runjeet Sing would not value the man's life at a pin's point, he would neglect no means of debarring him all access to food or air.

In the above quoted cases, except in those of Corfidius and Hermotinus, the absence of the spirit is alone suggested to the spectator by the condition of the body; since the memory of one state does not appear to have been carried into the other—if the spirit wandered into other regions it brings no tidings back; but we have many cases recorded where this deficient evidence seems to be supplied. The magicians and soothsayers of the northern countries, by narcotics, and other means, produce a cataleptic state of the body, resembling death, when their prophetic faculty is to be exercised; and although we all know that an alloy of imposition is generally mixed up with these exhibitions, still it is past a doubt, that a state of what we call clear-seeing is thus induced; and that on awaking, they bring tidings from various parts of the world of actions then performing and events occurring, which subsequent investigation have verified.

One of the most remarkable cases of this kind, is that recorded by Jung Stilling, of a man, who, about the year 1740 resided in the neighbourhood of Philadelphia, in the United States. His habits were retired, and he spoke little: he was grave, benevolent, and pious, and nothing was known against his character, except that he had the reputation of possessing some secrets that were not altogether *lawful*. Many extraordinary stories were told of him, and amongst the rest, the following:—The wife of a ship captain, whose husband was on a voyage to Europe and Africa, and from whom she had been long without tidings, overwhelmed with anxiety for his safety, was induced to address herself to this person. Having listened to her story, he begged he to excuse him for awhile, when he would bring her the intelligence she required. He then passed into an inner room, and she sat herself down to wait; but his absence continuing longer than she expected, she became impatient, thinking he had forgotten her; and so, softly approaching the door, she peeped through some aperture, and, to her surprise, beheld him lying on a sofa, as motionless as if he was dead. She, of course, did not think it advisable to disturb him, but waited his return, when he told her that her husband had not been able to write to her for such and such reasons; but that he was then in a coffee-house in London, and would very shortly be home again. Accordingly, he arrived, and as the lady learnt from him that the causes of his unusual silence had been precisely those alleged by the man, she felt extremely desirous of ascertaining the truth of the rest of the information; and in this she was gratified; for he no sooner set his eyes on the magician than he said that he had seen him before, on a certainday, in a coffee-house in London; and that he had told him that his wife was extremely uneasy about him; and that he, the captain, had thereon mentioned how he had been prevented writing; adding that he was on the eve of embarking for America. He had then lost sight of the stranger amongst the throng, and knew nothing more about him.

I have no authority for this story, but that of Jung Stilling; and if it stood alone, it might appear very incredible; but it is supported by so many parallel examples of information given by people in somnambulic states, that we are not entitled to reject it on the score of impossibility.

The late Mr. John Holloway, of the Bank of England, brother to the engraver of that name, related of himself that being one night in bed, with his wife, and unable to sleep, he had fixed his eyes and thoughts with uncommon intensity on a beautiful star that was shining in at the window, when he suddenly found his spirit released from his body and soaring into that bright sphere. But, instantly seized with anxiety for the anguish of his wife, if she discovered his body apparently dead beside her, he returned, and re-entered it with *difficulty* (*hence*, perhaps, the violent convulsions with which some somnambules of the highest order are awakened). He described that returning, was returning to darkness; and that whilst the spirit was free, he was *alternately in the light or the dark, accordingly as his thoughts were with his wife or with the star.* He said that he always avoided anything that could produce a repetition of this accident, the consequences of it being very distressing.

We know that by intense contemplation of this sort, the Dervishes produce a state of extacy, in which they pretend to be transported to other spheres; and not only the Seeress of Prevorst, but many other persons in a highly magnetic state, have asserted the same thing of themselves; and certainly the singular conformity of the intelligence they bring is not a little remarkable.

Dr. Kerner relates of his somnambule, Frederica Hauffe, that one day, at Weinsberg, she exclaimed in her sleep, " Oh ! God !" She immediately awoke, as if aroused by the exclamation, and said that she seemed to have heard two voices proceeding from herself. At this time, her father was lying dead in his coffin, at Oberstenfeld, and Dr. Fohr, the physician, who had attended him in his illness, was sitting with another person in an adjoining room, with the door open; when he heard the exclamation," Oh,God !" so distinctly, that, feeling certain there was nobody there, he hastened to the coffin, from whence the sound had appeared to proceed, thinking that Mr. W.'s death had been only apparent, and that he was reviving. The other person, who was an uncle of Frederica's, had heard nothing. No person was discovered from whom the exclamation could have proceeded, and the circumstance remained a mystery till an explanation ensued. Plutarch relates, that a certain man, called Thespesius, having fallen from a great height, was taken up apparently dead from the shock, although no external wound was to be discovered. On the third day after the accident, however, when they were about to bury him, he unexpectedly revived; and it was afterwards observed, to the surprise of all who knew him, that, from being a vicious reprobate, he became one of the most virtuous of men. On being interrogated with respect to the cause of the change, he related that, during the period of his bodily insensibility, it appeared to him that he was dead, and that he had been first plunged into the depths of an ocean, out of which, however, he soon emerged, and then, at one view, the whole of space was disclosed to him. Everything appeared in a different aspect, and the dimensions of the planetary bodies, and the intervals betwixt them, was tremendous; whilst his spirit seemed to float in a sea of light, like a ship in calm waters. He also described many other things that he had seen; he said that the souls of the dead, on quitting the body, appeared like a bubble of light, out of which a human form was quickly evolved. That, of these, some shot away at once in a direct line, with great rapidity, whilst others, on the contrary, seemed unable to find their due course, and continued to hover about, going hither and thither, till at length they also darted away in one direction or another. He recognized few of these persons he saw, but those whom he did, and sought to address, appeared as if they were stunned and amazed, and avoided him with terror. Their voices were indistinct, and seemed to be uttering vague lamentings. There were others, also, who floated farther from the earth, who looked bright, and were gracious; these avoided the approach of the last. In short, the demeanour and appearance of these spirits manifested clearly their degrees of joy or grief. Thespesius was then informed by one of them, that he was not dead, but that he had been permitted to come there by a divine decree, and that his soul, which was yet attached to his body, as by an anchor, would return to it again. Thespesius then observed, that he was different to the dead, by whom he was surrounded, and this observation seemed to restore him to his recollection. They were transparent, and environed by a radiance, but he seemed to trail after him a dark ray, or line of shadow. These spirits also presented very different aspects; some were entirely pervaded by a mild, clear, radiance, like that of the full moon; through others there appeared faint streaks, that diminished this splendour; whilst others, on the contrary, were distinguished by spots, or stripes of black, or of a dark colour, like the marks on the skin of a viper.

There is a circumstance which I cannot help here mentioning in connexion with this history of Thespesius, which on first reading it struck me very forcibly.

About three years ago, I had several opportunities of seeing two young girls, then under the care of a Mr. A., of Edinburgh, who hoped, chiefly by means of magnetism, to restore them to sight. One was a maid-servant afflicted with amaurosis, whom he had taken into his house from a charitable desire to be of use to her; the other, who had been blind from her childhood, was a young lady in better circumstances, the daughter of respectable tradespeople in the north of England. The girl with amaurosis was restored to sight, and the other was so far benefitted that she could distinguish houses, trees, carriages, &c., and, at length, though obscurely, the features of a person near her. At this period of the cure she was unhappily removed, and may possibly have relapsed into her former state. My reason, however, for alluding to these young women on this occasion, is, that they were in the habit of saying, when in the magnetic state—for they were both, more or less, *clairvoyantes*—that the people, whom Dr. A. was magnetising in the same room, presented very different appearances. Some of them they described as looking bright; whilst others were in different degrees, streaked with black.

One or two they mentioned over whom there seemed to hang a sort of cloud, like a ragged veil of darkness. They also said, though this was before any tidings of Baron von Reichenbach's discoveries had reached this country, that they saw light streaming from the fingers of Mr. A., when he magnetised them; and that sometimes his whole person seemed to them radiant. Now, I am positively certain that neither Mr. A., nor these girls, had ever heard of this story of Thespesius; neither had I, at that time; and I confess, when I did meet with it, I was a good deal struck by the coincidence. These young people said, that it was the "goodness or badness," meaning the moral state, of the persons that was thus indicated. Now, surely this concurrence betwixt the man, mentioned by Plutarch, and these two girls—the one of whom had no education whatever, and the other very little—is worthy of some regard.

I once asked a young person, in a highly clairvoyante state, whether she ever saw "the spirits of them that had passed away;" for so *she* designated the dead, never using the word *death* herself, in any of itsforms. She answered me, that she did.

"Then where are they?" I enquired.

"Some are waiting, and some are gone on before."

"Can you speak to them?" I asked.

"No," she replied, "there is no meddling nor no diretion."

In her waking state, she would have been quite incapable of these answers; and that "some are waiting and some gone on before," seems to be much in accordance with the vision of Thespesius.

Dr. Passavent mentions a peasant boy, who, after a short but painful illness, apparently died, his body being perfectly stiff. He, however, revived, complaining bitterly of being called back to life. He said he had been in a delightful place, and seen his deceased relations. There was a great exaltation of the faculties after this; and having been before rather stupid, he now, whilst his body lay stiff and immoveable and his eyes closed, prayed and discoursed with eloquence. He continued in this state for seven weeks, but finally recovered.

In the year 1733, Johann Schwerzeger fell into a similar state of trance, after an illness, but revived. He said he had seen his whole life, and every sin he had committed, even those he had quite forgotten—everything had been as present to him as when it happened. He also lamented being recalled from the happiness he was about to enter into; but said that he had only two days to spend in this valley of tears, during which time he wished everybody that would, should come and listen to what he had to tell them. His before sunken eyes now looked bright, his face had the bloom of youth, and he discoursed so eloquently that the minister said, they had exchanged offices, and the sick man had become his teacher. He died at the time he had foretold.

The most frightful cases of trance rcorded, are those in which the patient retains entire consciousness, although utterly unable to exhibit any evidence of life; and it is dreadful to think how many persons may have been actually buried, hearing every nail that was screwed into their own coffin, and as perfectly aware of the whole ceremony as those who followed them to the grave.

Dr. Binns mentions a girl, at Canton, who lay in this state, hearing every word that was said around her, but utterly unable to move a finger. She tried to cry out, but could not, and supposed that she was really dead. The horror of finding herself about to be buried, at length caused a perspiration to appear on her skin, and she finally revived. She de_ scribed that she felt that her soul had no power to act upon her body; and that it seemed to be *in her body and out of it, at the same time.*

Now, this is very much what the somnambulists say—their soul is out of the body, but is still so far in rapport with it, that it does not leave it entirely. Probably, magnetism would be the best means of reviving a person from this state.

The custom of burying people before there are unmistakable signs of death, is a very condemnable one. A Mr. M'G. fell into a trance, some few years since, and remained insensible for five days, his mother being, meanwhile, quite shocked that the physician would not allow him to be buried. He had, afterwards, a recurrence of the malady, which continued seven days.

A Mr. S., who had been some time out of the country, died, apparently, two days after his return. As he had eaten of a pudding which his step-mother had made for his dinner, with her own hands, people took into their heads she had poisoned him; and, the grave being opened for purposes of investigation, the body was found lying on its face.

One of the most frightful cases extant, is that of Dr. Walker, of Dublin, who had so strong a presentiment on this subject, that he had actually written a treatise against the Irish customs of hasty burial. He himself, subsequently died, as was believed, of a fever. His decease took place in the night, and on the following day he was interred. At this time, Mrs. Bellamy, the once celebrated actress, was in Ireland; and as she had promised him, in the course of conversation, that she would take care he should not be laid in the earth till unequivocal signs of dissolution had appeared, she no sooner heard of what had happened, than she took measures to have the grave re-opened; but it was, unfortunately, too late; Dr. Walker had evidently revived, and had turned upon his side; but life was now quite extinct. The case related by Lady Fanshawe, of her mother, is very remarkable, from the confirmation furnished by the event of her death.

"My mother, being sick of a fever," says Lady F., in her memoirs, "her friends and servants thought her deceased, and she lay in that state for two days and a night; but Mr. Winslow, coming to comfort my father, went into my mother's room, and, looking earnestly in her face, said, 'She was so handsome, and looked so lovely, that he could not think her dead; and, suddenly taking a lancet out of his pocket, he cut the sole of her foot, which bled, upon this he immediately caused her to be removed to the bed again, and to be rubbed, and such means used that she came to life, and opening her eyes, saw two of her kinswomen standing by her, Lady Knollys and Lady Russell, both with great wide sleeves, as the fashion then was; and she said, 'Did you not promise me fifteen years, and are you come again already?' which they, not understanding, bade her keep her spirits quiet in that great weakness wherein she was; but, some hours after, she desired my father and Dr. Howlesworth might be left alone with her, to whom she said, I will acquaint you, that, during my trance, I was in great grief, but in a place I could neither distinguish nor describe; but the sense of leaving my girl, who is dearer to me than all my children, remained a trouble upon my spirits. Suddenly I saw two by me, clothed in long white garments, and methought I fell down upon my face in the dust, and they asked me why I was so troubled in so great happiness. I replied, 'Oh, let me have the same grant given to Hezekiah, that I may live fifteen years to see my daughter a woman;' to which they answered, 'It is done!' and then at that instant I awoke out of my trance!' And Dr. Howlesworth did affirm, that that day she died, made just fifteen years from that time."

I have met with a somewhat similar case to this, which occurred to the mother of a very respectable person, now living in Edinburgh. She having been ill, was supposed to be dead, and preparations were making for her funeral, when, one of her fingers were seen to move, and, restoratives being applied, she revived As soon as she could speak, she said that she had been at the gates of heaven, where she saw some going in, but that they told her she was not ready. Amongst those who had passed her, and been admitted, she said, *she had seen Mr. So-and-so, the baker*, and the remarkable thing was, that during the time she had been in the trance, this man had died.

On the 10th of January, 1717, Mr. John Gardner, a minister, at Elgin, fell into a trance, and, being to all appearance dead, he was put into a coffin, and on the second day was carried to the grave. But fortunately a noise being heard, the coffin was opened, and he was found alive and taken home again; where, according to the record, "he related many strange and amazing things which he had seen in the other world."

Not to mention somnambules, there are numerous other cases recorded of persons who have said, on awaking from a trance, that they had been in the other world; though frequently the freed spirit, supposing that to be the interpretation of the mystery, seems busied with the affairs of the earth and brings tidings from distant places, as in the case of the American above-mentioned. Perhaps, in these latter cases, the disunion is less complete. Dr. Werner relates, of his somnambule, that it was after those attacks of catalepsy, in which her body had lain stiff and cold, that she used to say she had been wandering away through other spheres. Where the catalepsy is spontaneous and involuntary, and resembles death so nearly as not to be distinguished from it, we may naturally conclude, if we admit this hypothesis at all, that the seeing of the spirit would be clear in proportion to its disentanglement from the flesh.

I have spoken above of dream compelling or suggesting, and I have heard of persons who have a power of directing their own dreams to any particular subject.

This faculty may be, in some degree, analogous to that possessed by the American, and a few somnambulic persons, who appear to carry the recollections of one state into the other. The effects produced by the witch potions seem to have been somewhat similar, inasmuch as they dreamt what they expected or wished to dream. Jung Stilling mentions, that a woman gave in evidence, on a witch trial, that having visited the so-called witch, she had found her concocting a potion over the fire, of which she had advised her, the visitor, to drink, assuring her that she would then accompany her to the Sàbbath. The woman said, lest she should give offence, she had put the vessel to her lips, but had not drank of it; the witch, however, swallowed the whole, and immediately afterwards sunk down upon the hearth in a profound sleep, where she had left her. When she went to see her on the following day, she declared she had been to the Brocken.

Paolo Minucci relates, that a woman accused of sorcery, being brought before a certain magistrate, at Florence, she not only confessed her guilt, but she declared that, provided they would let her return home and anoint herself, she would attend the Sabbath that very night. The magistrate, a man more enlightened than the generality of his contemporaries, consented. The woman went home, used her unguent, and fell immediately into a profound sleep; whereupon they tied her to the bed, and tested the reality of the sleep by burns, blows, and pricking her with sharp instruments. When she awoke on the following day, she related that she had attended the Sabbath. I could quote several similar facts; and Gassendi actually endeavoured to undeceive some peasants who believed themselves witches, by composing an ointment that produced the same effects as their own magical applications.

In the year 1545, André Laguna, physician to Pope Julius III., anointed a patient of his, who was suffering from phrenzy and sleeplessness, with an unguent found in the house of a sorcerer, who had been arrested. The patient slept for thirty-six hours consecutively, and when, with much difficulty, she was awakened, she complained that they had torn her from the most ravishing delights; delights which seem to have rivalled the Heaven of the Mahometan. According to Llorente, the women who were dedicated to the service of the Mother of the Gods, heard continually the sounds of flutes and tambourines, beheld the joyous dances of the fauns and satyrs, and tasted of intoxicating pleasures, doubtless from a similar cause.

It is difficult to imagine, that all the unfortunate wretches who suffered death at the stake in the middle ages, for having attended the unholy assemblies they described, had no faith in their own stories; yet, in spite of the unwearied vigilance of public authorities, and private malignity, no such assemblage was ever detected. How, then, are we to account for the pertinacity of their confessions, but by supposing them the victims of some extraordinary delusion? In a paper addressed to the Inquisition, by Llorente, he does not scruple to assert, that the crimes imputed to, and confessed by, witches, have most frequently no existence but in their dreams; and that their dreams are produced by the drugs with which they anointed themselves.

The recipes for these compositions, which had descended traditionally from age to age, have been lost since witchcraft went out of fashion, and modern science has no time to investigate secrets which appear to be more curious than profitable; but in the profound sleep produced by these applications, it is not easy to say what phenomena may have occurred to justify, or, at least, account for, their self-accusations.

CHAPTER VII.

WRAITHS.

SUCH instances as that of Lady Fanshawe, and other similar ones, certainly seem to favour the hypothesis, that the spirit is freed from the body, when the latter becomes no longer a fit habitation for it. It does so when actual death supervenes, and the reason of its departure we may naturally conclude to be, that the body has ceased to be available for its manifestations; and in these cases, which seem so nearly allied to death, that, frequently, there would actually be no revival but for the exertions used, it does not seem very difficult to conceive that this separation may take place. When we are standing by a death bed, all we see is the death of the body, of the going forth of the spirit we see nothing; so in cases of apparent death, it may depart and return, whilst we are aware of nothing but the re-animation of the organism. Certain it is, that the Scriptures countenance this view of the case in several instances; thus, Luke says, Chap. viii., 34, "And he put them all out, and took her by the hand, and called, saying, 'Maid, arise!' And her spirit came again, and she arose straightway," &c., &c.

Dr. Wigan observes, when speaking of the effects of temporary pressure on the brain, that the mind is not annihilated, because, if the pressure is timely removed, it is restored, though, if continued too long, the body will be resolved into its primary elements; and he compares the human organism to a watch, which we can either stop or set going at will, which watch, he says, will also be gradually resolved into its ultimate elements by chemical action; and, he adds, that, to ask where the mind is, during the interruption, is like asking where the motion of the watch is. I think a wind instrument would be a better simile, for the motion of the watch is purely mechanical. It requires no informing, intelligent spirit to breathe into its apertures and make it the vehicle of the harshest discords, or of the most eloquent discourses. "The divinely mysterious essence, which we call the soul," he adds, "is not then the mind, from which it must be carefully distinguished, if we would hope to make any progress in mental philosophy. Where the soul resides during the suspension of the mental powers by asphyxia, I know not, any more than I know where it resided before it was united with that specific compound of bones, muscles, and nerve."

By a temporary pressure on the brain, the mind is certainly not annihilated, but its manifestations by means of the brain are suspended; the source of these manifestations being the soul or anima, in which dwells the life, fitting the temple for its divine inhabitant, the spirit. The connexion of the soul and the body is probably a much more intimate one than that of the latter with the spirit; though the soul, as well as the spirit, is immortal and survives when the body dies. Somnambulic persons seem to intimate that the soul of the fleshly body becomes, hereafter, the body of the spirit, as if the imago or idolon were the soul.

Dr. Wigan, and indeed physiologists in

general, do not appear to recognise the old distinction betwixt the pneuma or anima and the psyche—the soul and the spirit; and indeed the Scriptures occasionally seem to use the terms indifferently; but still there are passages enough which mark the distinction; as where St. Paul speaks of a "living soul and a quickening spirit," 1 Cor. xv., 45; again, 1 Thess. v., 23, "I pray God your whole spirit, and soul, and body, &c,;" and also, Hebrews iv. 12. Where he speaks of the sword of God "dividing asunder the soul and spirit." In Genesis, chap. ii., we are told that "man became a living soul;" but it is distinctly said, 1 Cor. xii., that the gifts of prophecy, the discerning of spirits, &c. &c. belong to the spirit. Then, with regard to the possibility of the spirit absenting itself from the body, St. Paul says, in referring to his own vision, 2 Cor. xii., "I knew a man in Christ, about fourteen years ago (whether in the body, I cannot tell; or out of the body, I cannot tell; God knoweth); such an one caught up to the third heaven;" and we are told, also, that to be "absent from the body is to be present with the Lord," and that when we are "at home in the body we are absent from the Lord." We are told, also, "the spirit returns to God, who gave it;" but it depends on ourselves whether or not our souls shall perish. We must suppose, however, that even in the worst cases some remnant of this divine spirit remains with the soul as long as the latter is not utterly perverted and rendered incapable of salvation.

St. John also says, that when he prophesied, he was in the *spirit;* but it was the "*Souls of the Slain*" that he saw, and that "cried with a loud voice, &c. &c." *Souls*, here, being probably used in the sense of individuals; as we say, "So many souls perished by shipwreck, &c."

In the "Revue de Paris," 29th July, 1838, it is related that a child *saw* the soul of a woman who was lying insensible in a magnetic crisis in which death nearly ensued, depart out of her; and I find recorded in another work that a somnambule who was brought to give advice to a patient, said, "It is too late; her soul is leaving her. I see the vital flame quitting her brain."

From some of the cases I have above related, we are led to the conclusion, that in certain conditions of the body, the spirit, in a manner unknown to us, resumes a portion of its freedom, and is enabled to exercise more or less of its inherent properties. It is somewhat released from those inexorable conditions of time and space, which bound and limit its powers, whilst in close connexion with matter, and it communes with other spirits who are also liberated. How far this liberation (if such it be), or re-integration of natural attributes, may take place in ordinary sleep, we can only conclude from examples. In prophetic dreams, and in those instances of information apparently received from the dead, this condition seems to occur; as, also, in such cases as that of the gentleman mentioned in a former chapter, who has several times been conscious on awaking, that he had been conversing with some one, whom he has been subsequently startled to hear had died at that period, and this is a man apparently in excellent health, endowed with a vigorous understanding, and immersed in active business.

In the story of the American, quoted in a former chapter from Jung Stilling, there was one point which I forbore to comment on at the moment, but to which I must now revert; this is the assertion, that the voyager had seen the man, and even conversed with him, in the coffee-house, in London, whence the desired intelligence was brought. Now, this single case standing alone, would amount to nothing, although Jung Stilling, who was one of the most conscientious of men, declares himself to have been quite satisfied with the authority on which he relates it; but, strange to say—for undoubtedly the thing *is* very strange—there are numerous similar instances recorded; and it seems to have been believed in all ages of the world, that people were sometimes seen, where bodily they were not; seen not by sleepers alone, but by persons in a perfect state of vigilance; and that this phenomenon, though more frequently occurring at the moment that the individual seen is at the point of death, does occasionally occur at indefinite periods anterior to the catastrophe; and sometimes where no such catastrophe is impending. In some of these cases, an earnest desire seems to be the cause of the phenomenon. It is not very long since a very estimable lady, who was dying in the Mediterranean, expressed herself perfectly ready to meet death, if she could but once more behold her children, who were in England. She soon afterwards fell into a comatose state, and the persons surrounding her were doubtful whether she had not already breathed her last; at all events, they did not expect her to revive. She did so, however, and now cheerfully announced that having seen her children, she was ready to depart. During the interval that she lay in this state, her family saw her in England, and were thus aware of her death before the intelligence reached them. As it is a subject, I understand, they are unwilling to speak of, I do not know precisely under what circumstances she was seen; but this is an exactly analogous case to that already recorded of Maria Goffe, of Rochester, who, when dying, away from home, expressed precisely the same feelings. She said she could not die happy till she had seen her children. By and by, she fell into a state of coma, which left them uncertain whether she was dead or alive. Her eyes were open and fixed, her jaw fallen, and there was no perceptible respiration. When she revived, she told her mother, who attended her, that she had been home and seen her children; which the other said was impossible, since she had been lying there in the bed the whole time. "Yes," replied the dying woman, "but I was there in my sleep." A widow woman, called Alexander, who had the care of these children, declared herself ready to take oath upon the sacrament, that during this period, she had seen the form of Maria Goffe come out of the room, where the eldest child slept, and approach the bed where she herself lay with the younger beside her. The figure had stood there nearly a quarter of an hour, as far as she could judge; and she remarked that the eyes and the mouth moved, though she heard no sound. She declared herself to have been perfectly awake, and that as it was the longest night in the year, it was quite light. She sat up in bed, and whilst she was looking on the figure, the clock on the bridge struck two. She then adjured the form in the name of God, whereupon it moved. She immediately arose and followed it; but could not tell what had become of it. She then became alarmed, and throwing on her clothes, went out and walked on the quay, returning to the house ever and anon to look at the children. At five o clock, she knocked at a neighbour's door, but they would not let her in. At six, she knocked again, and was then admitted, and related to them what she had seen, which they, of course, endeavoured to persuade her was a dream or an illusion. She declared herself, however, to have been perfectly awake; and said, that if ever she had seen Maria Goffe in her life, she had seen her that night.

The following story has been currently related in Rome, and is already in print. I take it from a German work, and I do not know how far its authenticity can be established. It is to the effect that two friends having agreed to attend confession together, one of them went at the appointed time to the Abbate B., and made his confession; after which the priest commenced the usual admonition, in the midst of which he suddenly ceased speaking. After waiting a short time, the penitent stept forward and perceived him lying in the confessional in a state of insensibility. Aid was summoned and means used to restore him, which were for some time ineffectual; at length, when he opened his eyes, he bade the penitent recite a prayer for his friend, who had just expired. This proved to be the case, on enquiry; and when the young man, who had naturally hastened to his friend's house, expressed a hope that he had not died without the last offices of the church, he was told to his amazement, that the Abbate B. had arrived just as he was in *extremis*, and had remained with him till he died.

These appearances seem to have taken place when the corporeal condition of the person seen elsewhere permits us to conceive the possibility of the spirit's having withdrawn from the body; but the question then naturally arises, what is it that was seen; and I confess, that of all the difficulties that surround the subject, I have undertaken to treat of, this seems to me the greatest; for we cannot suppose that a spirit can be visible to the human eye, and both in the above instances and several others I have to narrate, there is nothing that can lead us to the conclusion, that the persons who saw the wraith of double, were in any other than a normal state; the figure, in short, seems to have been perceived through their external organs of sense. Before I discuss this question, however, any further, I will relate some instances of a similar kind, only with this difference, that the wraith appearing as nearly as could be ascertained at the moment of death, it remains uncertain whether it was seen before or after the dissolution had taken place. As both in these cases above related and those that follow, the material body was visible in one place, whilst the wraith was visible in another, they appear to be strictly analogous; especially, as in both class of examples, the body itself was either dead or in a state that closely resembled death.

Instances of people being seen at a distance from the spot on which they are dying, are so numerous, that in this department I have positively an *embarras de richesses*, and find it difficult to make a selection; more especially as there is in each case little to relate, the whole phenomenon being comprised in the fact of the form being observed and the chief variations consisting in this, that the seer, or seers, frequently entertain no suspicion that what they have seen is any other than a form of flesh and blood; whilst on other occasions the assurance that the person is far away, or some peculiarity connected with the appearance itself, produces the immediate conviction that the shape is not corporeal.

Mrs. K., the sister of Provost B., of Aberdeen, was sitting one day with her husband, Dr. K., in the parlour of the manse, when she suddenly said, "Oh! there's my brother come! he has just passed the window," and, followed by her husband, she hastened to the door to meet the visitor. He was however not there. "He is gone round to the back door," said she; and thither they went; but neither was he there, nor had the servants seen any thing of him. Dr. K. said she must be mistaken; but she laughed at the idea; her brother had passed the window and looked in; he must have gone somewhere, and would doubtless be back directly. But he came not; and the intelligence shortly arrived from St. Andrew's, that at that precise time, as nearly as they could compare circumstances, he had died quite suddenly at his own place of residence. I have heard this story from connexions of the family, and also from an eminent professor of Glasgow, who told me that he had once asked Dr. K., whether he believed, in these appearances. "I cannot choose but believe," returned Dr. K., and then he accounted for his conviction by narrating the above particulars.

Lord and Lady M. were residing on their estate in Ireland; Lord M. had gone out shooting in the morning; and was not expected to return till towards dinner time. In the course of the afternoon, Lady M. and a friend were walking on the terrace that forms a promenade in front of the castle, when she said, "Oh, there is M. returning!" whereupon she called to him to join them. He, however, took no notice, but walked on before them, till they saw him enter the house, whither they followed him; but he was not to be found; and before they had recovered their surprise at his sudden disappearance, he was brought home dead; having been killed by his own gun. It is a curious fact in this case, that whilst the ladies were walking behind the figure, on the terrace, Lady M. called the attention of her companion to the shooting jacket, observing that it was a particularly convenient one, and that she had the credit of having contrived it for him herself.

A person in Edinburgh, busied about her daily work, saw a woman enter her house with whom she was on such ill terms that she could not but be surprized at the visit; but whilst she was expecting an explanation, and under the influence of her resentment avoiding to look at her, she found she was gone. She remained quite unable to account for the visit, and as she said, "Was wondering what had brought her there," when she heard that the woman had expired at that precise time.

Madame O. B. was engaged to marry an officer who was with his regiment in India; and wishing to live in privacy till the union took place, she retired to the country and boarded with some ladies of her acquaintance, awaiting his return. She, at length, heard that he had obtained an appointment, which, by improving his prospects, had removed some difficulties out of the way of the marriage, and that he was immediately coming home. A short time after the arrival of this intelligence, this lady and one of those with whom she was residing, were walking over a bridge, when the friend said, alluding to an officer, she saw on the other side of the way, " what an extraordinary expression of face." But without pausing to answer, Madame O. B. darted across the road to meet the stranger—but he was gone ! Where ? They could not conceive. They ran to the toll-keepers at the ends of the bridge to enquire if they had observed such a person ; but they had not. Alarmed and perplexed, for it was her intended husband that she had seen, Madame O. B. returned home ; and in due time the packet that should have brought himself, brought the sad tidings of his unexpected death.

Madame O. B. never recovered the shock, and died herself of a broken heart not long afterwards.

Mr. H., an eminent artist, was walking arm in arm, with a friend, in Edinburgh, when he suddenly left him, saying, " Oh, there's my brother !" He had seen him with the most entire distinctness, but was confounded by losing sight of him, without being able to ascertain whither he had vanished. News came, ere long, that at that precise period his brother had died.

Mrs. T., sitting in her drawing-room, saw her nephew, then at Cambridge, pass across the adjoining room. She started up to meet him, and, not finding him, summoned the servants to ask where he was. They, however, had not seen him, and declared he could not be there; whilst she as positively declared he was. The young man had died, at Cambridge, quite unexpectedly.

A Scotch minister went to visit a friend, who was dangerously ill. After sitting with the invalid for some time, he left him to take some rest, and went below. He had been reading in the library some little time, when, on looking up, he saw the sick man standing at the door. "God bless me!" he cried, starting up, "how can you be so imprudent?" The figure disappeared; and hastening upstairs, he found his friend had expired.

Three young men, at Cambridge, had been out hunting, and afterwards dined together in the apartments of one of them. After dinner, two of the party, fatigued with their morning's exercise, fell asleep, whilst the third, a Mr. M., remained awake. Presently the door opened, and a gentleman entered and placed himself behind the sleeping owner of the rooms, and, after standing there a minute, proceeded into the gyp-room—a small inner chamber, from which there was no egress.

Mr. M. waited a little while, expecting the stranger would come out again ; but as he did not, he awoke his host, saying, "There's somebody gone into your room ; I don't know who it can be."

The young man rose and looked into the gyp-room, but there being nobody there he naturally accused Mr. M. of dreaming ; but the other assured him he had not been asleep. He then described the stranger—an elderly man, &c. &c. dressed like a country squire, with gaiters on, and so forth. "Why that's my father," said the host, and he immediately made enquiry, thinking it possible the old gentleman had slipt out unobserved by Mr. M. He was not, however, to be heard of ; and the post shortly brought a letter announcing that he had died at the time he had been seen in his son's chamber at Cambridge.

Mr. C. F. and some young ladies were not long ago, standing together looking in at a shop window, at Brighton, when he suddenly darted across the way and they saw him hurrying along the street, apparently in pursuit of somebody. After waiting a little while, as he did not return, they went home without him ; and when he come, they of course arraigned him for his want of gallantry.

"I beg your pardon," said he; "but I saw an acquaintance of mine that owes me some money, and I wanted to get hold of him."

"And did you?" enquired the ladies.

"No," returned he; "I kept sight of him some time; but I suddenly missed him. I can't think how."

No more was thought of the matter; but by the next morning's post, Mr. C. F. received a letter, enclosing a draught from the father of the young man he had seen, saying, that his son had just expired; and that one of his last requests had been that he would pay Mr. C. F. the money that he owed him.

Two young ladies staying at the Queen's Ferry, arose one morning early, to bathe; as they descended the stairs, they each exclaimed, "There's my uncle!" They had seen him standing by the clock. He died at that time.

Very lately, a gentleman living in Edinburgh, whilst sitting with his wife, suddenly arose from his seat, and advanced towards the door, with his hand extended, as if about to welcome a visitor. On his wife's enquiring what he was about, he answered that he had seen so-and-so enter the room. She had seen nobody. A day or two afterwards the post brought a letter announcing the death of the person seen.

A regiment, not very long since, stationed at New Orleans, had a temporary mess-room erected, at one end of which was a door for the officers; and at the other, a door and a space railed off for the messman. One day, two of the officers were playing at chess, or draughts, one sitting with his face towards the centre of the room, the other with his back to it. "Bless me! why, surely that is your brother!" exclaimed the former to the latter, who looked eagerly round, his brother being then, as he believed, in England. By this time, the figure having passed the spot where the officers were sitting, presented only his back to them. "No," replied the second, "that is not my brother's regiment; that's the uniform of the Rifle Brigade. By heavens! it *is* my brother, though;" he added, starting up, and eagerly pursuing the stranger, who at that moment turned his head and looked at him, and then, somehow, strangely disappeared amongst the people standing at the messman's end of the room. Supposing he had gone out that way, the brother pursued him, but he was not to be found; neither had the messman, nor any body there, observed him. The young man died at that time in England, having just exchanged into the Rifle Brigade.

I could fill pages with similar instances, not to mention those recorded in other collections and in history. The case of Lord Balcarres is perhaps worth alluding to, from its being so perfectly well established. Nobody has ever disputed the truth of it, only they get out of the difficulty by saying that it was a spectral illusion! Lord B. was in confinement in the castle of Edinburgh, under suspicion of Jacobitism, when one morning, whilst lying in bed, the curtains were drawn aside by his friend, Viscount Dundee, who looked upon him steadfastly, leaned for some time on the mantlepiece and then walked out of the room. Lord B. not supposing that what he saw was a spectre, called to Dundee to come back and speak to him, but he was gone; and shortly afterwards the news came that he had fallen about that same hour at Killicranky.

Finally, I have met with three instances of persons who are so much the subjects of this phenomenon, that they see the wraith of most persons that dies belonging to them, and frequently of those who are merely acquaintance. They see the person as if he were alive, and unless they know him positively to be elsewhere, they have no suspicion but that it is himself, in the flesh, that is before them, till the sudden disappearance of the figure brings the conviction. Sometimes, as in the case of Mr. C. F. above alluded to, no suspicion arises, till the news of the death arrives, and they mention, without reserve, that they have met so and so, but he did not stop to speak, and so forth.

On other occasions, however, the circumstances of the appearance are such, that the seer is instantly aware of its nature. In the first place, the time and locality may produce the conviction.

Mrs. J. wakes her husband in the night, and tells him she has just seen her father pass through the room—she being in the West Indies and her father in England. He died that night. Lord T. being at sea, on his way to Calcutta, saw his wife enter his cabin.

Mrs. Mac...., of Sky, went from Lynedale where she resided, to pay a visit in Perthshire. During her absence, there was a ball given at L.; and when it was over, three young ladies, two of them her daughters, assembled in their bedroom to talk over the evening's amusement. Suddenly, one of them cried, "O God! my mother." They all saw her pass across the room towards a chest of drawers, where she vanished. They immediately told their friends what they had seen; and afterwards learnt that the lady died that night.

Lord M. being from home, saw Lady M., whom he had left two days before, perfectly well, standing at the foot of his bed; aware of the nature of the appearance, but wishing to satisfy himself that it was not a mere spectral illusion, he called his servant, who slept in the dressing-room, and said to him, "John, who's that?" "It's my lady!" replied the man. Lady M. had been seized with inflammation and died after a few hours illness. This circumstance awakened so much interest at the time, that I am informed by a member of the family, George the Third was not satisfied without hearing the particulars both from Lord M. and the servant, also.

But, besides time and locality, there are very frequently other circumstances accompanying the appearance, which not only show the form to be spectral, but also make known to the seer the nature of the death that has taken place.

A lady, with whose family I am acquainted, had a son abroad. One night she was lying in bed, with a door open which led into an adjoining room, where there was a fire. She had not been to sleep, when she saw her son cross this adjoining room and approach the fire, over which he leant, as if very cold. She saw that he was shivering and dripping wet. She immediately exclaimed, "That's my G.!" The figure turned its face round, looked at her sadly, and disappeared. That same night the young man was drowned.

Mr. P., the American manager, in one of his voyages to England, being in bed, one night, between sleeping and waking, was disturbed by somebody coming into his cabin, dripping with water. He concluded that the person had fallen overboard, and asked him why he came there to disturb him, when there were plenty of other places for him to go to? The man muttered something indistinctly, and Mr. P. then perceived that it was his own brother. This roused him completely, and feeling quite certain that somebody had been there, he got out of bed to feel if the carpet was wet on the spot where his brother stood. It was not, however; and when he questioned his shipmates, the following morning, they assured him that nobody had been overboard, nor had anybody been in his cabin. Upon this, he noted down the date and the particulars of the event, and, on his arrival at Liverpool, sent the paper sealed, to a friend in London, desiring it might not be opened till he wrote again. The Indian post, in due time, brought the intelligence that on that night Mr. P.'s brother was drowned.

A similar case to this is that of Captain Kidd, which Lord Byron used to say he heard from Captain K himself. He was, one night awakened in his hammock, by feeling something heavy lying upon him. He opened his eyes, and saw, or thought he saw, by the indistinct light in the cabin, his brother, in uniform, lying across the bed. Concluding that this was only an illusion arising out of some foregone dream, he closed his eyes again to sleep; but again he felt the weight, and there was the form still lying across the bed. He now stretched out his hand, and felt the uniform, which was quite wet. Alarmed, he called out for somebody to come to him; and, as one of the officers entered, the figure disappeared. He afterwards learnt, that his brother was drowned on that night in the Indian Ocean.

Ben Jonson told Drummond, of Hawthornden, that, being at Sir Robert Cotton's house, in the country, with old Cambden, he saw, in a vision, his eldest son, then a child at London, appear to him with a mark of a bloody cross on his forehead; at which, amazed, he prayed to God; and, in the morning, mentioned the circumstance to Mr. Cambden, who persuaded him it was fancy. In the mean time, came letters announcing that the boy had died of the plague. The custom of indicating an infected house by a red cross, is here suggested; the cross, apparently, symbolizing the manner of the death.

Mr. S. C. a gentleman of fortune, had a son in India. One fine calm summer's morning, in the year 1780, he and his wife were sitting at breakfast, when she arose and went to the window; upon which, turning his eyes in the same direction, he started up and followed her, saying, "My dear, do you see that?" "Surely," she replied, "it is our son. Let us go to him!" As she was very much agitated, however, he begged her to sit down and recover herself; and when they looked again, the figure was gone. The appearance was that of their son, precisely as they had last seen him. They took note of the hour, and afterwards learnt that he had died in India at that period.

A lady, with whose family I am acquainted, was sitting with her son, named Andrew, when she suddenly exclaimed that she had seen him pass the window, in a white mantle. As the window was high from the ground, and overhung a precipice, no one could have passed; else, she said, "Had there been a path, and he not beside her at the moment, she should have thought he had walked by on stilts." Three days afterwards, Andrew was seized with a fever which he had caught from visiting some sick neighbours; and expired after a short illness.

In 1807, when several people were killed in consequence of a false alarm of fire, at Sadler's Wells, a woman named Price, in giving her evidence at the inquest, said, that her little girl had gone into the kitchen about half-past ten o'clock, and was surprised to see her brother there, whom she supposed to be at the Theatre. She spoke to him; whereupon, he disappeared. The child immediately told her mother, who, alarmed, set off to the theatre and found the boy dead.

In the year 1813, a young lady in Berlin, whose intended husband was with the army at Dusseldorf, heard some one knock at the door of her chamber, and her lover entered in a white negligé, stained with blood. Thinking that this vision proceeded from some disorder in herself, she arose and quitted the room to call the servant; who not being at hand, she returned, and found the figure there still. She now became much alarmed, and having mentioned the circumstance to her father, enquiries were made of some prisoners that were marching through the town, and it was ascertained, that the young man had been wounded and had been carried to the house of Dr. Ehrlick, in Leipsick, with great hopes of recovery. It afterwards proved, however, that he had died at that period, and that his last thoughts were with her. This lady earnestly wished and prayed for another such visit; but she never saw him again.

In the same year, a woman in Bavaria, who had a brother with the army in Russia, was one day at field-work, on the skirts of a forest, and everything quiet around her, when she repeatedly felt herself hit by small stones, though, on looking round, she could see nobody. At length, supposing it was some jest, she threw down her implements and stept into the wood whence they had proceeded, when she saw a headless figure, in a soldier's mantle, leaning against a tree. Afraid to approach, she summoned some labourers from a neighbouring field, who also saw it; but on going up to it, it disappeared. The woman declared her conviction that the circumstance indicated her brother's death; and it was afterwards ascertained that he had, on that day, fallen in a trench.

Some few years ago, a Mrs. H., residing in Limerick, had a servant whom she much esteemed, called Nelly Hanlon. Nelly was a very steady person, who seldom asked for a holiday, and consequently Mrs. H. was the less disposed to refuse her, when she requested a day's leave of absence for the purpose of attending a fair, that was to take place a few miles off. The petition was therefore favorably heard, but when Mr. H. came home and was informed of Nelly's proposed excursion, he said she could not be spared, as he had invited some people to dinner for that day, and he had nobody he could trust with the keys of the cellar except Nelly; adding, that it was not likely his business would allow him to get home time enough to bring up the wine himself.

Unwilling, however, after giving her consent, to disappoint the girl, Mrs. H. said that she would herself undertake the cellar department on the day in question; so when the wished for morning arrived, Nelly departed in great spirits, having faithfully promised to return that night, if possible, or at the latest, the following morning.

The day passed as usual and nothing was thought about Nelly, till the time arrived for fetching up the wine, when Mrs. H, proceeded to the cellar stairs with the key, followed by a servant carrying a bottle-basket. She had, however, scarcely begun to descend when she uttered a loud scream and dropt down in a state of insensibility. She was carried up stairs and laid upon the bed, whilst, to the amazement of the other servants, the girl who had accompanied her, said, that they had seen Nelly Hanlon, dripping with water, standing at the bottom of the stairs. Mr. H. being sent for, or coming home at the moment, this story was repeated to him; whereupon he reproved the woman for her folly; and, proper restoratives being applied, Mrs. H. at length began to revive. As she opened her eyes, she heaved a deep sigh saying, " Oh, Nelly Hanlon," and as soon as she was sufficiently recovered to speak, she corroborated what the girl had said; she had seen Nelly at the foot of the cellar stairs, dripping as if she had just come out of the water. Mr. H. used his utmost efforts to persuade his wife out of what he looked upon to be an illusion; but in vain. " Nelly," said he, " will come home by and by and laugh at you," whilst she, on the contrary, felt sure that Nelly was dead.

The night came, and the morning came, but there was no Nelly. When two or three days had passed, enquiries were made; and it was ascertained that she had been seen at the fair, and had started to return home in the evening; but from that moment all traces of her were lost, till her body was ultimately found in the river. How she came by her death, was never known. Now, in most of these cases, which I have above detailed, the person was seen where his dying thoughts might naturally be supposed to have flown, and the visit seems to have been made either immediately before or immediately after the dissolution of the body; in either case we may imagine that the final parting of the spirit had taken place, even if the organic life was not quite extinct. I have met with some cases in which we are not left in any doubt, with respect to what were the last wishes of the dying person: for example,—a lady, with whom I am acquainted, was on her way to India, when near the end of her voyage, she was one night awakened by a rustling in her cabin, and a consciousness that there was something hovering about her. She sat up, and saw a bluish cloudy form moving away; but persuading herself it must be fancy, she addressed herself again to sleep; but as soon as she lay down, she both heard and felt the same thing: it seemed to her as if this cloudy form hung over and enveloped her. Overcome with horror, she screamed. The cloud then moved away, assuming distinctly a human shape. The people about her naturally persuaded her that she had been dreaming; and she wished to think so; but when she arrived in India, the first thing she heard was, that a very particular friend had come down to Calcutta to be ready to receive her on her landing, but that he had been taken ill and died, saying, he only wished to live to see his old friend once more. He had expired on the night she saw the shadowy form in her room.

A very frightful instance of this kind of phenomenon is related by Dr. H. Werner, of Baron Emilius von O. This young man had been sent to prosecute his studies in Paris; but forming some bad connexions, he became dissipated, and neglected them. His father's counsels were unheeded, and his letters remained unanswered. One day the young baron was sitting alone on a seat, in the Bois de Boulogne, and had fallen somewhat into a reverie, when, on raising his eyes, he saw his father's form before him. Believing it to be a mere spectral illusion, he struck at the shadow with his riding-whip, upon which it disappeared. The next day brought him a letter urging his return home instantly, if he wished to see his parent alive. He went, but found the old man already in his grave. The persons who had been about him said, that he had been quite conscious, and had a great longing to see his son; he had, indeed, exhibited one symptom of delirium, which was, that after expressing this desire, he had suddenly exclaimed, "My God! he is striking at me with his riding-whip!" and immediately expired. In this case, the condition of the dying man resembles that of a somnambulist, in which the patient describes what he sees taking place at a distance; and the archives of magnetism furnish some instances, especially that of Auguste Müller, of Karlsruhe, in which, by the force of will, the sleeper has not only been able to bring intelligence from a distance, but also, like the American magician, to make himself visible. The faculties of prophecy and clear or far-seeing, frequently disclosed by dying persons, is fully acknowledged by Dr. Abercrombie, and other physiologists.

Mr. F. saw a female relative, one night, by his bed-side. Thinking it was a trick of some one to frighten him, he struck at the figure; whereon she said, "What have I done? I know I should have told it you before." This lady was dying at a distance, earnestly desiring to speak to Mr. F. before she departed.

I will conclude this chapter with the following extract from "Lockhart's Life of Scott":—

"Walter Scott to Daniel Terry, April 30, 1818. (The new house at Abbotsford being then in progress, Scott living in an older part, close adjoining.)

" * * The exposed state of my house has led to a mysterious disturbance. The night before last we were awakened by a violent noise, like drawing heavy boards along the new part of the house. I fancied something had fallen, and thought no more about it. This was about *two* in the morning. Last night, at the same witching hour, the very same noise occurred. Mrs. S., as you know, is rather timbersome; so up I got, with Beardie's broad sword under my arm—

"Bolt upright,
And ready to fight."

But nothing was out of order, neither can I discover what occasioned the disturbance.

Mr. Lockhart adds, "On the morning that Mr. Terry received the foregoing letter, in London, Mr. William Erskine was breakfasting with him, and the chief subject of their conversation was the sudden death of George Bullock, which had occurred on the same night, and, as nearly as they could ascertain, at the very hour when Scott was roused from his sleep by the 'mysterious disturbance' here described. This coincidence, when Scott received Erskine's minute detail of what had happened in Tenterdon-street (that is the death of Bullock, who had the charge of furnishing the new rooms at Abbotsford), made a much stronger impression on his mind than might be gathered from the tone of an ensuing communication."

It appears that Bullock had been at Abbotsford, and made himself a great favourite with old and young. Scott, a week or two afterwards, wrote thus to Terry, "Were you not struck with the fantastical coincidence of our nocturnal disturbances at Abbotsford, with the melancholy event that followed? I protest to you, the noise resembled half-a-dozen men hard at work, putting up boards and furniture; and nothing can be more certain than that there was nobody on the premises at the time. With a few additional touches the story would figure in Glanville or Aubrey's collection. In the mean time, you may set it down with poor Dubisson's warnings, as a remarkable coincidence coming under your own observation."

CHAPTER VIII.

DÖPPELGANGERS, OR DOUBLES.

In the instances detailed in the last chapter the apparition has shown itself, as nearly as could be discovered, at the moment of dissolution; but there are many cases in which the wraith is seen at an indefinite period before or after the catastrophe. Of these, I could quote a great number, but as they generally resolve themselves into simply seeing a person where they were not, and death ensuing very shortly afterwards, a few will suffice.

There is a very remarkable story of this kind, related by Macnish, which he calls "a case of hallucination, arising without the individual being conscious of any physical cause by which it might be occasioned." If this case stood alone, strange as it is, I should think so, too; but when similar instances abound, as they do, I cannot bring myself to dispose of it so easily. The story is as follows:—Mr. H. was one day walking along the street, apparently in perfect health, when he saw, or supposed he saw, his acquaintance, Mr. C., walking before him. He called to him, aloud, but he did not seem to hear him, and continued moving on. Mr. H. then quickened his pace for the purpose of overtaking him, but the other increased his, also, as if to keep ahead of his pursuer, and proceeded at such a rate that Mr. H. found it impossible to make up to him. This continued for some time, till, on Mr. C. reaching a gate, he opened it and passed in, slamming it violently in Mr. H.'s face. Confounded at such treatment from a friend, the latter instantly opened the gate, and looked down the long lane into which it led, where, to his astonishment, no one was to be seen. Determined to unravel the mystery, he then went to Mr. C.'s house, and his surprise was great to hear that he was confined to his bed, and had been so for several days. A week or two afterwards, these gentlemen met at the house of a common friend, when Mr. H. related the circumstance, jocularly telling Mr. C. that, as he had seen his wraith, he of course could not live long. The person addressed, laughed heartily, as did the rest of the party; but in a few days, Mr. C. was attacked with putrid sore throat, and died; and, within a short period of his death, Mr. H. was also in the grave.

This is a very striking case: the hastening on and the actually opening and shutting the gate, evincing not only *will* but *power* to produce mechanical effects, at a time the person was bodily elsewhere. It is true he was ill, and, it is highly probable, was at the time asleep. The showing himself to Mr. H., who was so soon to follow him to the grave, is another peculiarity which appears frequently to attend these cases, and which seems like what was in old English, and is still, in Scotch, called a *tryst*—an appointment to meet again betwixt those spirits, so soon to be free. Supposing Mr. C. to have been asleep, he was possibly, in that state, aware of what impended over both.

There is a still more remarkable case, given by Mr. Barham, in his reminiscences. I have no other authority for it; but he relates, as a fact, that a respectable young woman was awaked, one night, by hearing somebody in her room, and that on looking up, she saw a young man, to whom she was engaged. Extremely offended by such an intrusion, she bade him instantly depart, if he wished her ever to speak to him again. Whereupon, he bade her not be frightened; but said he was come to tell her that he was to die that day six weeks, and then disappeared. Having ascertained that the young man himself could not possibly have been in her room, she was naturally much alarmed, and, her evident depression leading to some enquiries, she communicated what had occurred to the family with whom she lived—I think as dairy-maid; but I quote from memory. They attached little importance to what seemed so improbable, more especially as the young man continued in perfectly good health, and entirely ignorant of this prediction, which his mistress had the prudence to conceal from him. When the fatal day arrived, these ladies saw the girl looking very cheerful, as they were going for their morning's ride, and observed to each other that the prophecy did not seem likely to be fulfilled; but when they returned, they saw her running up the avenue towards the house, in great agitation, and learned that her lover was either dead, or dying, I think, in consesequence of an accident.

The only key I can suggest as the explanation of such a phenomenon as this, is, that the young man, in his sleep, was aware of the fate that awaited him; and that whilst his body lay in his bed, in a state approaching to trance or catalepsy, the freed spirit—free as the spirits of the actual dead—went forth to tell the tale to the mistress of his soul.

Franz von Baader, says in a letter to Dr. Kerner, that Eckartshausen, shortly before his death, assured him that he possessed the power of making a person's double or wraith appear, whilst his body lay elsewhere, in a state of trance or catalepsy. He added that the experiment might be dangerous, if care were not taken to prevent intercepting the rapport of the etherial form with the material one.

A lady, an entire disbeliever in these spiritual phenomena, was one day walking in her own garden with her husband, who was indisposed, leaning on her arm, when seeing a man with his back towards them, and a spade in his hand, digging, she exclaimed, "Look there! Who's that?" "Where?" said her companion;

and at that moment, the figure leaning on the spade, turned round, and looked at her, sadly shaking its head; and she saw it was her husband. She avoided an explanation, by pretending she had made a mistake. Three days afterwards the gentleman died; leaving her entirely converted to a belief she had previously scoffed at.

Here, again, the foreknowledge and evident design, as well as the power of manifesting it, is extremely curious. More especially, as the antitype of the figure was neither in a trance nor asleep, but perfectly conscious, walking and talking. If any particular purpose were to be gained, by the information indicated, the solution might be less difficult. One object, it is true, may have been, and indeed, was attained, namely, the change in the opinions of the wife; and it is impossible to say, what influence such a conversion may have had on her after life.

It must be admitted that these cases are very perplexing. We might, indeed, get rid of them by denying them, but the instances are too numerous, and the phenomenon has been too well known in all ages to be set aside so easily. In the above examples the apparition, or wraith, has been in some way connected with the death of the person whose visionary likeness is seen; and, in most of these instances, the earnest longing to behold those beloved, seems to have been the means of effecting the object. The mystery of death is to us so awful and impenetrable, and we know so little of the mode in which the spiritual and the corporeal are united and kept together during the continuance of life, or what condition may ensue when this connexion is about to be dissolved, that whilst we look with wonder upon such phenomena as these above alluded to, we yet find very few persons who are disposed to reject them as utterly apocryphal. They feel that in that department, already so mysterious, there may exist a greater mystery still; and the very terror with which the thoughts of present death inspires most minds, deters people from treating this class of facts with that scornful scepticism with which many approximate ones are denied and laughed at. Nevertheless, if we suppose the person to have been dead, though it be but an inappreciable instant of time, before he appears, the appearance comes under the denomination of what is commonly called a ghost; for, whether the spirit has been parted from the body one second or fifty years, ought to make no difference in our appreciation of the fact, nor is the difficulty less in one case than the other.

I mention this, because I have met with, and do meet with, people constantly, who admit this class of facts, whilst they declare they cannot believe in ghosts; the instances, they say, of people being seen at a distance at the period of their death, are too numerous to permit of the fact being denied. In granting it, however, they seem to me to grant everything. If, as I have said above, the person be dead, the form seen is a ghost or spectre, whether he has been dead a second or a century; if he be alive, the difficulty is certainly not diminished, on the contrary, it appears to me to be considerably augmented; and it is to this perplexing class of facts I shall next proceed; namely, those in which the person is not only alive, as in some of the cases above related, but where the phenomenon seems to occur without any reference to the death of the subject, present or prospective.

In either case, we are forced to conclude that the thing seen is the same; the questions are, what is it that we see, and how does it render itself visible; and, still more difficult to answer, appears the question, of how it can communicate intelligence, or exert a mechanical force. As, however, this investigation will be more in its place when I have reached that department of my subject commonly called ghosts, I will defer it for the present, and merely confine myself to that of Doubles, or Döppelgangers, as the Germans denominate the appearance of a person out of his body.

In treating of the case of Auguste Muller, a remarkable somnambule, who possessed the power of appearing elsewhere, whilst his body lay cold and stiff in his bed. Professor Kieser, who attended him, says, that the phenomenon, as regards the seer, must be looked upon as purely subjective—that is, that there was no outstanding form of Auguste Muller visible to the sensuous organs, but that the magnetic influence of the somnambule, by the force of his will, acted on the imagination of the seer, and called up the image which he believed he saw. But then, allowing this to be possible, as Dr. Werner says, how are we to account for those numerous cases in which there is no somnambule concerned in the matter, and no especial rapport, that we are aware of, established betwixt the parties? And yet these latter cases are much the most frequent; for, although I have met with numerous instances recorded by the German physiologists of what is called far-working on the part of their somnambules, this power of appearing out of the body seems to be a very rare one. Many persons will be surprised at these allusions to a kind of magnetic phenomena, of which, in this country, so little is known or believed; but the physiologists and psychologists of Germany have been studying this subject for the last fifty years, and the volumes filled with their theoretical views and records of cases, are numerous beyond anything the English public has an idea of.

The only other theory I have met with, which pretends to explain the mode of this double appearance, is that of the spirit leaving the body, as we have supposed it to do in cases of dreams and catalepsy; in which instances, the nerve-spirit, which seems to be the archæus or astral spirit of the ancient philosophers, has the power of projecting a visible body out of the imponderable matter of the atmosphere. According to this theory, this nerve-spirit, which seems to be an embodiment of—or rather, a body constructed out of the nervous fluid, or ether—in short, the spiritual body of St. Paul, is the bond of union betwixt the body and the soul, or spirit; and has the plastic force of raising up an aerial form. Being the highest organic power, it cannot by any other, physical or chemical, be destroyed ; and when the body is cast off, it follows the soul; and as, during life, it is the means by which the soul acts upon the body, and is thus enabled to communicate with the external world, so when the spirit is disembodied, it is through this nerve-spirit, that it can make itself visible, and even exercise mechanical powers.

It is certain, that not only somnambules, but sick persons, are occasionally sensible of a feeling that seems to lend some countenance to this latter theory.

The girl at Canton, for example, mentioned in a former chapter, as well as many somnambulic patients, declare, whilst their bodies are lying stiff and cold, that they see it, as if out of it; and, in some instances, they describe particulars of its appearance, which they could not see in the ordinary way. There are also numerous cases of sick persons seeing themselves double, where no tendency to delirium or spectral illusion had been observed. These are, in this country, always placed under the latter category ; but I find various instances recorded by the German physiologists, where this appearance has been seen by others, and even by children, at the same time that it was *felt* by the invalid. In one of these cases, I find the sick person saying, "I cannot think, how I am lying. It seems to me that I am divided and lying in two places at once." It is remarkable, that a friend of my own, during an illness in the autumn of 1845, expressed precisely the same feeling; we however, saw nothing of this second *ego;* but it must be remembered, that the seeing these things, as I have said in a former chapter, probably depends on a peculiar faculty or condition of the seer. The servant of Elisha was not blind, but yet he could not see what his master saw, till his eyes were opened—that is, till he was rendered capable of perceiving spiritual objects.

When Peter was released from prison by the angel—and it is not amiss here to remark, that even he "wist not that it was true which was done by the angel, but thought he saw a vision," that is, he did not believe his senses, but supposed himself the victim of a spectral illusion—but when he was released, and went and knocked at the door of the gate, where many of his friends were assembled, they not conceiving it possible he could have escaped, said, when the girl who had opened the door, insisted that he was there, "It is his angel." What did they mean by this? The expression is not *an* angel, but *his* angel. Now, it is not a little remarkable, that in the East, to this day, a double, or döppleganger, is called a man's angel, or messenger. As we cannot suppose that this term was used otherwise than seriously by the disciples that were gathered together in Mark's house, for they were in trouble about Peter, and when he arrived were engaged in prayer, we are entitled to believe that they alluded to some recognized phenomenon. They knew, either that the likeness of a man—his spiritual self—sometimes appeared where bodily he was not; and that this imago or idolon was capable of exerting a mechanical force, or else that other spirits sometimes assumed a mortal form, or they would not have supposed it to be Peter's angel that had *knocked* at the gate.

Dr. Ennemoser, who always leans to the physical, rather than the psychical explanation of a phenomenon, says, that the faculty of self-seeing, which is analogous to seeing another person's double, is to be considered an illusion; but that this imago of another seen at a distance, at the moment of death, must be supposed to have an objective reality.

But if we are capable of thus perceiving the imago of another person, I cannot comprehend why we may not see our own ; unless, indeed, the former was never perceived, but when the body of the person seen, was in a state of insensibility ; but this does not always seem to be a necessary condition, as will appear by some examples I am about to detail. The faculty of perceiving the object, Dr. Ennemoser considers analogous to that of second sight, and thinks it may be evolved by local, as well as idiosyncratical, conditions. The difficulty arising from the fact, that some persons are in the habit of seeing the wraiths of their friends and relations must be explained by his hypothesis. The spirit, as soon as liberated from the body, is adapted for communion with *all* spirits; embodied or otherwise, but all embodied spirits are not prepared for communion with it.

A Mr. R., a gentleman who has attracted public attention by some scientific discoveries, had had a fit of illness at Rotterdam. He was in a state of convalescence, but was still so far taking care of himself as to spend part of the day in bed, when, as he was lying there one morning, the door opened, and there entered, in tears, a lady with whom he was intimately acquainted, but whom at the time he believed to be in England. She walked hastily up to the side of his bed, wrung her hands, evincing by her gestures extreme anguish of mind, and before he could sufficiently recover his surprise to enquire the cause of her distress and sudden appearance, she was gone. She did not disappear, but walked out of the room again, and Mr. R. immediately summoned the servants of the hotel, for the purpose of making enquiries about the English lady—when she came, what had happened to her, and where she had gone to, on quitting his room? The people declared there was no such person there; he insisted there was, but they at length convinced him that they, at least, knew nothing about her. When his physician visited him, he naturally expressed the great perplexity into which he had been thrown by this circumstance: and, as the doctor could find no symptoms about his patient that could warrant a suspicion of spectral illusion, they made a note of the date and hour of the occurrence, and Mr. R. took the earliest opportunity of ascertaining if anything had happened to the lady in question. Nothing had happened to herself, but at that precise period her son had expired, and she was actually in the state of distress in which

Mr. R. beheld her. It would be extremely interesting to know whether her thoughts had been very intensely directed to Mr. R. at the moment; but that is a point which I have not been able to ascertain. At all events, the impelling cause of the form projected, be the mode of it what it may, appears to have been violent emotion. The following circumstance, which is forwarded to me by the gentleman to whom it occurred, appears to have the same origin:—

"On the evening of the 12th of March, 1792," says Mr. H., an artist, and a man of science, "I had been reading in the 'Philosophical Transactions,' and retired to my room somewhat fatigued, but not inclined to sleep. It was a bright moonlight night, and I had extinguished my candle and was sitting on the side of the bed, deliberrately taking off my clothes, when I was amazed to behold the visible appearance of my half-uncle, Mr. R. Roberston, standing before me; and, at the same instant, I heard the words, '*Twice will be sufficient!*' The face was so distinct that I actually saw the pock-pits. His dress seemed to be made of a strong twilled sort of sackcloth, and of the same dingy colour. It was more like a woman's dress than a man's—resembling a petticoat, the neck-band close to the chin, and the garment covering the whole person, so that I saw neither hands nor feet. Whilst the figure stood there, I twisted my fingers till they cracked, that I might be sure I was awake.

"On the following morning, I enquired if anybody had heard lately of Mr. R., and was well laughed at when I confessed the origin of my enquiry. I confess I thought he was dead; but when my grandfather heard the story, he said that the dress I described, resembled the straight-jacket Mr. R. had been put in formerly, under an attack of insanity. Subsequently, we learnt that on the night, and at the very hour I had seen him, he had attempted suicide, and been actually put into a straight-jacket.

"He afterwards recovered, and went to Egypt with Sir Ralph Abercrombie. Some people laugh at this story, and maintain that it was a delusion of the imagination; but surely this is blinking the question! Why should my imagination create such an image, whilst my mind was entirely engrossed with a mathematical problem?"

The words "*Twice will be sufficient*," probably embodied the thought, uttered or not, of the maniac, under the influence of his emotion —two blows or two stabs would be sufficient for his purpose.

Dr. Kerner relates a case of a Dr. John B., who was studying medicine in Paris, seeing his mother, one night, shortly after he had got into bed, and before he had put out his light. She was dressed after a fashion in which he had never seen her; but she vanished; and thus aware of the nature of the appearance, he became much alarmed, and wrote home to enquire after her health. The answer he received was, that she was extremely unwell, having been under the most intense anxiety on his account, from hearing that several medical students in Paris had been arrested as resurrectionists; and, knowing his passion for anatomical investigations, she had apprehended he might be amongst the number. The letter concluded with an earnest request that he would pay her a visit. He did so, and his surprise was so great on meeting her, to perceive that she was dressed exactly as he had seen her in his room at Paris, that he could not, at first, embrace her, and was obliged to explain the cause of his astonishment and repugnance.

An analogous case to these is that of Dr.

Donne, which is already mentioned in so many publications, that I should not allude to it here, but for the purpose of showing that these examples belong to a *class* of facts, and that it is not to be supposed that similarity argues identity, or that one and the same story is reproduced with new names and localities. I mention this, because when circumstances of this kind are related, I sometimes hear people say, "Oh, I have heard that story before, but it was said to have happened to Mr. So-and-so, or at such a place; the truth being, that these things happen in all places, and to a great variety of people.

Dr. Donne was with the embassy, in Paris, where he had been but a short time, when his friend Mr. Roberts entering the *salon*, found him in a state of considerable agitation. As soon as he was sufficiently recovered to speak, he said that his wife had passed twice through the room, with a dead child in her arms. An express was immediately dispatched to England to enquire for the lady, and the intelligence returned was, that, after much suffering, she had been delivered of a dead infant. The delivery had taken place at the time that her husband had seen her in Paris. Nobody has ever disputed Dr. Donne's assertion that he saw his wife, but, as usual, the case is crammed into the theory of spectral illusions. They say, Dr. Donne was naturally very anxious about his wife's approaching confinement, of which he must have been aware; and that his excited imagination did all the rest. In the first place, I do not find it recorded that he was suffering any particular anxiety on the subject; and even if he were, the coincidences in time and in the circumstance of the dead child, remain unexplained. Neither are we led to believe that the doctor was unwell, or living the kind of life that is apt to breed thick-coming fancies. He was attached to the embassy in the gay city of Paris; he had just been taking luncheon with others of the *suite*, and had been left alone but a short time, when he was found in the state of amazement above described. If such extraordinary cases of spectral illusion as this, and many others I am recording, can suddenly arise in constitutions apparently healthy, it is certainly high time that the medical world reconsider the subject, and give us some more comprehensible theory of it; if they are not cases of spectral illusion, but are to be explained under that vague and abused term *Imagination*, let us be told something more about Imagination—a service which those who consider the word sufficient to account for these strange phenomena, must, of course, be qualified to perform. If, however, both these hypotheses—for they are but simple hypotheses, unsupported by any proof whatever, only being delivered with an air of authority in a rationalistic age, they have been allowed to pass unquestioned—if, however, they are not found sufficient to satisfy a vast number of minds, which I know to be the case, I think the enquiry I am instituting cannot be wholly useless or unacceptable, let it lead us where it may. The *truth* is all I seek; and I think there is a very important truth to be educed from the further investigation of this subject in its various relations—in short, a truth of paramount importance to all others; one which contains evidence of a fact, in which we are more deeply concerned than in any other; and which, if well established, brings demonstration to confirm intuition and tradition. I am very well aware of all the difficulties in the way—difficulties internal and external; many inherent to the subject itself; and others extraneous, but inseparable from it; and I am very far from supposing that my book is to settle the question, even with a single mind. All I hope or expect is, to show that the question is not disposed of yet, either by the rationalists or the physiologists; and that it is still an open one; and all I desire is, to arouse enquiry and curiosity; and that thus some mind, better qualified than mine, to follow out the investigation, may be incited to undertake it.

Dr. Kerner mentions the case of a lady, named Dillenius, who was awakened one night by her son, a child of six years of age; her sister-in-law, who slept in the same room, also awakened at the same time, and all three saw Madame Dillenius enter the room, attired in a black dress, which she had lately bought. The sister said, "I see you double! you are in bed, and yet you are walking about the room." They were both extremely alarmed, whilst the figure stood between the doors, in a melancholy attitude, with the head leaning on the hand. The child, who also saw it, but seems not to have been terrified, jumped out of bed, and running to the figure, put his hand through it as he attempted to push it, exclaiming, "Go away, you black woman." The form, however, remained as before; and the child, becoming alarmed, sprung into bed again. Madame Dillenius expected that the appearance foreboded her own death; but that did not ensue. A serious accident immediately afterwards occurred to her husband, and she fancied there might be some connexion betwixt the two events.

This is one of those cases that, from their extremely perplexing nature, have induced some psychologists to seek an explanation in the hypothesis, that other spirits may for some purpose or under certain conditions, assume the form of a person with a view to giving an intimation or impression, which the gulf separating the material from the spiritual world renders it difficult to convey. As regards such instances as that of Madame Dillenius, however, we are at a loss to discover any motive—unless, indeed, it be sympathy—for such an exertion of power, supposing it to be possessed; but in the famous case of Catherine of Russia, who is said, whilst lying in bed, to have been seen by the ladies to enter the throne-room and being informed of the circumstance, went herself and saw the figure seated on the throne, and bade her guards fire on it, we may conceive it possible that her guardian spirit, if such she had, might adopt this mode of warning her to prepare for a change, which, after such a life as hers, we are entitled to conclude, she was not very fit to encounter.

There are numerous examples of similar phenomena to be met with. Professor Stilling relates that he heard from the son of a Madame M., that his mother, having sent her maid up stairs, on an errand, the woman came running down in a great fright, saying that her mistress was sitting above, in her arm-chair, looking precisely as she had left her below. The lady went up stairs, and saw herself as described by the woman, very shortly after which she died.

Dr. Werner relates, that a jeweller at Ludwigsburg, named Ratzel, when in perfect health, one evening, on turning the corner of a street, met his own form, face to face; the figure seemed as real and life-like as himself; and he was so close as to look into its very eyes. He was seized with terror, and it vanished. He related the circumstance to several people, and endeavoured to laugh, but, nevertheless, it was evident he was painfully impressed with it. Shortly afterwards, as he was passing through a forest, he fell in with some wood-cutters, who asked him to lend a hand to the ropes with which they were pulling down an oak tree. He did so, and was killed by its fall.

Becker, professor of mathematics at Rostock, having fallen into argument with some friends, regarding a disputed point of theology, on going to his library to fetch a book which he wished to refer to, saw himself sitting at the table in the seat he usually occupied. He approached the figure, which appeared to be reading, and, looking over its shoulder, he observed that the book open before it was a Bible, and that, with one of the fingers of the right hand, it pointed to the passage, "Make ready thy house, for thou must die." He returned to the company, and related what he had seen, and, in spite of all their arguments to the contrary, remained fully persuaded that his death was at hand. He took leave of his friends, and expired on the following day, at six o'clock in the evening. He had already attained a considerable age. Those who would not believe in the appearance, said he had died of the fright; but, whether he did so or not, the circumstance is sufficiently remarkable; and, if this were a real, outstanding apparition, it would go strongly to support the hypothesis alluded to above, whilst, if it were a spectral illusion, it is, certainly, an infinitely strange one.

As I am aware how difficult it is, except where the appearance is seen by more persons than one, to distinguish cases of actual self-seeing from those of spectral illusion, I do not linger longer in this department, but, returning to the analogous subject of Doppelgangers, I will relate a few curious instances of this kind of phenomenon.

Stilling relates, that a Government officer, of the name of Triplin, in Weimar, on going to his office to fetch a paper of importance, saw his own likeness sitting there, with the deed before him. Alarmed, he returned home, and desired his maid to go there and fetch the paper she would find on the table. The maid saw the same form, and imagined that her master had gone by another road, and got there before her; his mind seems to have preceded his body.

The Landrichter, or Sheriff F., in Frankfort, sent his secretary on an errand; presently afterwards, the secretary re-entered the room, and laid hold of a book. His master asked him what had brought him back, whereupon the figure vanished, and the book fell to the ground, it was a volume of Linnæus. In the evening, when the secretary returned, and was interrogated with regard to his expedition, he said that he had fallen into an eager dispute with an acquaintance, as he went along, about some botanical question, and had ardently wished he had had his Linnæus with him to refer to.

Dr. Werner relates, that Professor Happach had an elderly maid-servant, who was in the habit of coming every morning to call him, and on entering the room, which he generally heard her do, she usually looked at a clock which stood under the mirror. One morning she entered so softly that though he saw her, he did not hear her foot; she went, as was her custom, to the clock, and came to his bedside, but suddenly turned round and left the room. He called after her, but she not answering, he jumped out of bed and pursued her. He could not see her, however, till he reached her room, where he found her fast asleep in bed. Subsequently, the same thing occurred frequently with this woman.

An exactly parallel case was related to me as occurring to himself, by a publisher in Edinburgh. His housekeeper was in the habit of calling him every morning. On one occasion, being perfectly awake, he saw her enter, walk to the window, and go out again without speaking. Being in the habit of fastening his door, he supposed he had omitted to do so ; but presently afterwards he heard her knocking to come in, and he found the door was still locked. She assured him she had not been there before. He was in perfectly good health at the time this happened.

Only a few nights since, a lady, with whom I am intimately acquainted, was in bed, and had not been to sleep, when she saw one of her daughters, who slept in an upper room, and who had retired to rest some time before, standing at the foot of her bed. "H—," she said, "what is the matter? what are you come for?" The daughter did not answer, but moved away. The mother jumped out of bed, but not seeing her, got in again: but the figure was still there. Perfectly satisfied it was really her daughter, she spoke to her, asking if anything had happened ; but again the figure moved silently away, and again the mother jumped out of bed, and actually went part of the way up stairs; and this occurred a third time. The daughter was during the whole of this time asleep in her bed ; and the lady herself is quite in her usual state of health ; not robust, but not by any means sickly, nor in the slightest degree hysterical or nervous; yet, she is perfectly convinced that she saw the figure of her daughter on that occasion, though quite unable to account for the circumstance. Probably the daughter was dreaming of the mother.

Edward Stern, author of some German works, had a friend, who was frequently seen *out of the body*, as the Germans term it; and the father of that person was so much the subject of this phenomenon, that he was frequently observed to enter his house, whilst he was yet working in the fields. His wife used to say to him, "Why, papa, you came home before;" and he would answer, "I dare say; I was so anxious to get away earlier, but it was impossible."

The cook in a convent of nuns, at Ebersdorf, was frequently seen picking herbs in the garden, when she was in the kitchen and much in need of them.

A Danish Physician, whose name Dr. Werner does not mention, is said to have been frequently seen entering a patient's room, and on being spoken to, the figure would disappear, with a sigh. This used to occur when he had made an appointment which he was prevented keeping, and was rendered uneasy by the failure. The hearing of it, however, occasioned him such an unpleasant sensation that he requested his patients never to tell him when it happened.

A president of the Supreme Court, in Ulm, named Pfizer, attests the truth of the following case:—A gentleman, holding an official situation, had a son at Gottingen, who wrote home to his father, requesting him to send him, without delay, a certain book, which he required to aid him in preparing a dissertation he was engaged in. The father answered, that he had sought but could not find the work, in question. Shortly afterwards, the latter had been taking a book from his shelves, when, on turning round, he beheld, to his amazement, his son just in the act of stretching up his hand towards one on a high shelf in another part of the room. "Hallo!" he exclaimed, supposing it to be the young man himself; but the figure disappeared; and, on examining the shelf, the father found there the book that was required, which he immediately forwarded to Gottingen; but before it could arrive there, he received a letter from his son, describing the exact spot where it was to be found.

A case of what is called spectral illusion

is mentioned by Dr. Paterson, which appears to me to belong to the class of phenomena I am treating of. One Sunday evening, Miss N. was left at home, the sole inmate of the house, not being permitted to accompany her family to church, on account of her delicate state of health. Her father was an infirm old man, who seldom went from home, and she was not aware whether, on this occasion, he had gone out with the rest or not. By and by, there came on a severe storm of thunder, lightning, and rain, and Miss N. is described as becoming very uneasy about her father. Under the influence of this feeling, Dr. Paterson says, she went into the back room, where he usually sat, and there saw him in his arm chair. Not doubting but it was himself, she advanced, and laid her hand upon his shoulder, but her hand encountered vacancy; and, alarmed, she retired. As she quitted the room, however, she looked back, and there still sat the figure. Not being a believer in what is called the "supernatural," Miss N. resolved to overcome her apprehensions, and return into the room, which she did, and saw the figure as before. For the space of fully half an hour she went in and out of the room in this manner, before it disappeared. She did not see it vanish, but the fifth time she returned, it was gone; Dr. Paterson vouches for the truth of this story, and no doubt of its being a mere illusion occurs to him, though the lady had never before or since, as she assured him, been troubled with the malady. It seems to me much more likely that, when the storm came on, the thoughts of the old man would be intensely drawn homewards, he would naturally wish himself in his comfortable arm-chair, and knowing his young daughter to be alone, he would inevitably feel some anxiety about her, too. There was a mutual projection of their spirits towards each other; and the one that was most easily freed from its bonds, was seen where in the spirit it actually was; for, as I have said above, a spirit out of the flesh, to whom space is annihilated, must be where its thoughts and affections are, for its thoughts and affections are *itself*.

I observe that Sir David Brewster, and others, who have written on this subject, and who represent all these phenomena as images projected on the retina from the brain, dwell much on the fact that they are seen alike, whether the eye be closed or open. There are, however, two answers to be made to this argument; first, that even if it were so, the proof would not be decisive; since it is generally with closed eyes that somnambulic persons see —whether natural somnambules or magnetic patients; and, secondly, I find in some instances which appear to me to be genuine cases of an objective appearance, that where the experiment has been tried, the figure is not seen when the eyes are closed.

The author of a work, entitled "An Inquiry into the Nature of Ghosts," who adopts the illusion theory, relates the following story, as one he can vouch for, though not permitted to give the names of the parties:—

"Miss —, at the age of seven years, being in a field not far from her father's house, in the parish of Kirklinton, in Cumberland, saw what she thought was her father in the field, at a time that he was in bed, from which he had not been removed for a considerable period. There were in the field, also, at the same moment, George Little, and John, his fellow-servant. One of these cried out, "Go to your father, Miss!" She turned round, and the figure had disappeared. On returning home, she said, "Where is my father?" The mother answered, "In bed, to be sure, child;" out of which he had not been.

I quote this case, because the figure was seen by two persons; I could mention several similar instances, but when only seen by one, they are, of course, open to another explanation.

Goethe, whose family, by the way, were ghost-seers, relates, that as he was once in an uneasy state of mind, riding along the foot-path towards Drusenheim, he saw, "not with the eyes of his body, but with those of his spirit," himself on horseback coming towards him, in a dress that he then did not possess. It was grey, and trimmed with gold; the figure disappeared; but eight years afterwards he found himself, quite accidentally, on that spot, on horseback, and in precisely that attire." This seems to have been a case of *second sight*. The story of Byron's being seen in London when he was lying in a fever at Patras, is well known; but may possibly have arisen from some extraordinary personal resemblance, though so firm was the conviction of its being his actual self that a bet of a hundred guineas was offered on it.

Some time ago, the "Dublin University Magazine" related a case, I know not on what authority, as having occurred at Rome, to the effect, that a gentleman had, one night on going home to his lodging, thrown his servant into great amazement—the man exclaiming, "Good Lord, sir! you came home before!" He declared that he had let his master into the house, attended him up stairs, and, I think, undressed him, and seen him get into bed. When they went to the room, they found no clothes; but the bed appeared to have been lain in, and there was a strange mark upon the ceiling, as if from the passage of an electrical fluid. The only thing the young man could remember, whereby to account for this extraordinary circumstance was, that whilst abroad, and in company, he had been overcome with ennui, fallen into a deep reverie, and had for a time forgotten that he was not at home.

When I read this story, though I have learnt from experience to be very cautious how I pronounce that impossible which I know nothing about, I confess it somewhat exceeded my receptive capacity, but I have since heard of a similar instance, so well authenticated, that my incredulity is shaken.

Dr. Kerner relates, that a canon of a catholic cathedral, of somewhat dissipated habits, on coming home one evening, saw a light in his bed-room. When the maid opened the door, she started back with surprise, whilst he enquired why she had left a candle burning up stairs; upon which she declared, that he had come home just before, and gone to his room, and she had been wondering at his unusual silence. On ascending to his chamber, he saw himself sitting in the arm-chair. The figure arose, passed him, and went out at the room-door. He was extremely alarmed, expecting his death was at hand. He, however, lived many years afterwards, but the influence on his moral character was very beneficial.

Not long since, a professor, I think of theology, at a college at Berlin, addressed his class, saying, that, instead of his usual lecture, he should relate to them a circumstance which, the preceding evening, had occurred to himself, believing the effects would be no less salutary.

He then told them that, as he was going home the last evening, he had seen his own imago, or double, on the other side of the street. He looked away, and tried to avoid it, but, finding it still accompanied him, he took a short cut home, in hopes of getting rid of it, wherein he succeeded, till he came opposite his own house, when he saw it at the door.

It rang, the maid opened, it entered, she handed it a candle, and, as the professor stood in amazement, on the other side of the street, he saw the light passing the windows, as it wound its way up to his own chamber. He then crossed over and rang; the servant was naturally dreadfully alarmed on seeing him, but, without waiting to explain, he ascended the stairs. Just as he reached his own chamber, he heard a loud crash, and, on opening the door, they found no one there, but the ceiling had fallen in, and his life was thus saved. The servant corroborated this statement to the students; and a minister, now attached to one of the Scotch churches, was present when the professor told his tale. Without admitting the doctrine of protecting spirits, it is difficult to account for these latter circumstances.

A very interesting case of an apparent friendly intervention occurred to the celebrated Dr. A. T., of Edinburgh. He was sitting up late one night, reading in his study, when he heard a foot in the passage, and knowing the family were, or ought to be, all in bed, he rose and looked out to ascertain who it was, but, seeing nobody, he sat down again. Presently, the sound recurred, and he was sure there was somebody, though he could not see him. The foot, however, evidently ascended the stairs, and he followed it, till it led him to the nursery door, which he opened, and found the furniture was on fire; and thus, but for this kind office of his good angel, his children would have been burnt in their beds.

The most extraordinary history of this sort, however, with which I am acquainted, is the following, the facts of which are perfectly authentic:—

Some seventy or eighty years since, the apprentice, or assistant, of a respectable surgeon in Glasgow, was known to have had an illicit connexion with a servant girl, who somewhat suddenly disappeared. No suspicion, however, seems to have been entertained of foul play. It appears rather to have been supposed that she had retired for the purpose of being confined, and, consequently, no enquiries were made about her.

Glasgow was, at that period, a very different place to what it is at present, in more respects than one; and, amongst its peculiarities, was the extraordinary strictness with which the observance of the Sabbath was enforced, insomuch, that nobody was permitted to show themselves in the streets or public walks during the hours dedicated to the church services; and there were actually inspectors appointed to see that this regulation was observed, and to take down the names of defaulters.

At one extremity of the city, there is some open ground, of rather considerable extent, on the north side of the river, called "The Green," where people sometimes resort for air and exercise; and where lovers not unfrequently retire to enjoy as much solitude as the proximity to so large a town can afford.

One Sunday morning, the inspectors of public piety above alluded to having traversed the city, and extended their perquisitions as far as the lower extremity of the Green, where it was bounded by a wall, observed a young man lying on the grass, whom they immediately recognized to be the surgeon's assistant. They, of course, enquired why he was not at church, and proceeded to register his name in their books, but, instead of attempting to make any excuse for his offence, he only rose from the ground, saying, "I am a miserable man; look in the water!" He then immediately crossed a style, which divided the wall and led to a path extending along the side of the river towards the Rutherglen-road. They saw him cross the style, but, not comprehending the significance of his words, instead of observing him further, they naturally directed their attention to the water, where they presently perceived the body of a woman. Having with some difficulty dragged it ashore, they immediately proceeded to carry it into the town, assisted by several other persons, who by this time had joined them. It was now about one o'clock, and, as they passed through the streets, they were obstructed by the congregation that was issuing from one of the principal places of worship; and, as they stood up for a moment, to let them pass, they saw the surgeon's assistant issue from the church door. As it was quite possible for him to have gone round some other way, and got there before them, they were not much surprised. He did not approach them, but mingled with the crowd, whilst they proceeded on their way.

On examination, the woman proved to be the missing servant-girl. She was pregnant, and had evidently been murdered with a surgeon's instrument, which was found entangled amongst her clothes. Upon this, in consequence of his known connexion with her, and his implied self-accusation to the inspectors, the young man was apprehended on suspicion of being the guilty party, and tried upon the circuit. He was the last person seen in her company, immediately previous to her disappearance; and there was, altogether, such strong presumptive evidence against him, as corroborated by what occurred on the green would have justified a verdict of *guilty*. But, strange to say, this last most important item in the evidence failed, and he established an incontrovertible *alibi;* it being proved, beyond all possibility of doubt, that he had been in church from the beginning of the service to the end of it. He was, therefore, acquitted; whilst the public were left in the greatest perplexity, to account as they could for this extraordinary discrepancy. The young man was well known to the inspectors, and it was in broad daylight that they had met him and placed his name in their books. Neither, it must be remembered, were they seeking for him, nor thinking of him, nor of the woman, about whom there existed neither curiosity nor suspicion. Least of all, would they have sought her where she was, but for the hint given to them.

The interest excited, at the time, was very great; but no natural explanation of the mystery has ever been suggested.

CHAPTER IX.

APPARITIONS.

THE number of stories on record, which seem to support the views I have suggested in my last chapter, is, I fancy, little suspected by people in general; and still less is it imagined that similar occurrences are yet frequently taking place. I had, indeed, myself no idea of either one circumstance or the other, till my attention being accidentally turned in this direction, I was led into enquiries, the result of which has extremely surprised me. I do not mean to imply that all my acquaintance are ghost-seers, or that these things happen every day; but the amount of what I do mean, is this: first, that besides the numerous instances of such phenomena alluded to in history, which have been treated as fables by those who profess to believe the rest of the narratives, though the whole rests upon the same foundation, *i. e.*, tradition and hear-say; besides these, there exists in one form or another, hundreds and hundreds of recorded cases, in all countries, and in all languages, exhibiting that degree of similarity which mark them as belonging to a class of facts, many of these being of a nature which seems to preclude the possibility of bringing them under the theory of spectral illusions; and, secondly, that I scarcely meet any one man or woman, who, if I can induce them to believe I will not publish their names, and am not going to laugh at them, is not prepared to tell me of some occurrence of the sort, as having happened to themselves, their family, or their friends. I admit that in many instances they terminate their narration, by saying, that they think it must have been an illusion, *because* they cannot bring themselves to believe in ghosts; not unfrequently adding, that they *wish* to think so; since to think otherwise, would make them uncomfortable.

I confess, however, that this seems to me a very unwise, as well as a very unsafe way of treating the matter. Believing the appearance to be an illusion, *because* they cannot bring themselves to believe in ghosts, simply amounts to saying, "I don't believe, because I don't believe;" and is an argument of no effect, except to invalidate their capacity for judging the question, at all; but the second reason for not believing, namely, that they do not wish to do so, has not only the same disadvantage, but is liable to much more serious objections; for it is our duty to ascertain the truth in an affair that concerns every soul of us so deeply; and to shrink from looking at it, lest it should disclose something we do not like, is an expedient as childish as it is desperate. In reviewing my late novel of "Lilly Dawson," where I announce the present work, I observe, that, whilst some of the reviewers scout the very idea of any body's believing in ghosts, others, less rash, whilst they admit that it is a subject we know nothing about, object to further investigation, on account of the terrors and uncomfortable feelings that will be engendered. Now, certainly, if it were a matter in which we had no personal concern, and which belonged merely to the region of speculative curiosity, every body would be perfectly justified in following their inclinations with regard to it; there would be no reason for frightening themselves, if they did not like it; but since it is perfectly certain that the fate of these poor ghosts, be what it may, will be ours some day—perhaps before another year or another week has passed over our heads—to shut our eyes to the truth, because it may, perchance, occasion us some uncomfortable feelings, is surely a strange mixture of contemptible cowardice and daring temerity. If it be true that by some law of nature, departed souls occasionally revisit the earth, we may be quite certain that it was intended we should know it, and that the law is to some good end; for no law of God can be purposeless or mischievous; and is it conceivable that we should say, we will not know it, because it is disagreeable to us? Is not this very like saying, "Let us eat, drink, and be merry, for tomorrow we die!" and yet refusing to enquire what is to become of us when we do die? refusing to avail ourselves of that demonstrative proof, which God has mercifully placed within our reach? And with all this obstinacy, people do not get rid of the apprehension; they go on struggling against it and keeping it down by argument and reason, but there are very few persons indeed, men or women, who, when placed in a situation, calculated to suggest the idea, do not feel the intuitive conviction striving within them. In the ordinary circumstances of life, nobody suffers from this terror; in the extraordinary ones, I find the professed disbelievers not much better off than the believers. Not long ago, I heard a lady expressing the great alarm she should have felt, had she been exposed to spend a whole night on Ben Lomond, as Margaret Fuller, the American authoress, did lately; "for," said she, "though I don't believe in ghosts, I should have been dreadfully afraid of seeing one, then!"

Moreover, though I do not suppose that man, in his normal state, could ever encounter an incorporeal spirit without considerable awe, I am inclined to think that the extreme terror the idea inspires, arises from bad training. The ignorant frighten children with ghosts; and the better educated assure them there is no such thing. Our understanding may believe the latter, but our instincts believe the former; so that, out of this education, we retain the terror, and just belief enough to make it very troublesome whenever we are placed in circumstances that awaken it. Now, perhaps, if the thing were differently managed, the result might be different. Suppose the subject were duly investigated, and it were ascertained that the views I and many others are disposed to entertain with regard to it, are correct; and suppose, then, children were calmly told that it is not impossible, but that on some occasion they may see a departed friend again; that the laws of nature established by an allwise Providence, admit of the dead sometimes revisiting the earth, doubtless for the benevolent purpose of keeping alive in us our faith in a future state; that death is merely a transition to another life, which it depends on ourselves to make happy or otherwise; and that, whilst those spirits which appear bright and blessed, may well be objects of our envy, the others should excite only our intense compassion. I am persuaded that a child so educated would feel no terror at the sight of an apparition, more especially as there very rarely appears to be anything terrific in the aspect of these forms; they generally come in their "habits as they lived," and appear so much like the living person in the flesh, that where they are not known to be already dead, they are frequently mistaken for them. There are exceptions to this rule, but it is very rare that the forms in themselves exhibit anything to create alarm.

As a proof that a child would not naturally be terrified at the sight of an apparition, I will adduce the following instance, the authenticity of which I can vouch for:—

A lady with her child embarked on board a vessel at Jamaica, for the purpose of visiting her friends in England, leaving her husband behind her quite well. It was a sailing packet; and they had been some time at sea, when, one evening, whilst the child was kneeling before her, saying his prayers, previously to going to rest, he suddenly said, looking eagerly to a particular spot in the cabin, "Mamma, Papa!" "Nonsense, my dear!" the mother answered; "You know your papa is not here!" "He is, indeed, mama," returned the child, "he is looking at us now!" Nor could she convince him to the contrary. When she went on deck, she mentioned the circumstance to the captain, who thought it so strange, that he said he would note down the date of the occurrence. The lady begged him not to do so, saying, it was attaching a significance to it which would make her miserable; he did it, however, and shortly after her arrival in England, she learnt that her husband had died exactly at that period.

I have met with other instances in which children have seen apparitions without exhibiting any alarm; and in the case of Fredericka Hauffe, the infant in her arms was frequently observed to point smilingly to those which she herself said she saw. In the above related case, we find a valuable example of an apparition which we cannot believe to have been a mere subjective phenomenon, being seen by one person and not by another. The receptivity of the child may have been greater, or the rapport betwixt it and its father stronger, but this occurrence inevitably leads us to suggest, how often our departed friends may be near us, and we not see them!

A Mr. B., with whom I am acquainted, informed me that some years ago, he lost two children. There was an interval of two years between their deaths; and about as long a period had elapsed since the decease of the second, when the circumstance I am about to relate took place. It may be conceived that at that distance of time, however vivid the impression had been at first, it had considerably faded from the mind of a man engaged in business; and he assures me that on the night this event occurred, he was not thinking of the children at all; he was, moreover, perfectly well, and had neither eaten or drank anything unusual, nor abstained from eating or drinking anything to which he was accustomed. He was, therefore, in his normal state; when shortly after he had lain down in bed, and before he had fallen asleep, he heard the voice of one of the children say, "Papa! Papa!"

"Do you hear that?" he said to his wife, who lay beside him; "I hear Archy calling me, as plain as ever I heard him in my life!"

"Nonsense!" returned the lady; "you are fancying it."

But presently he again heard "Papa! Papa!" and now both voices spoke. Upon which, exclaiming "I can stand this no longer!" he started up, and drawing back the curtains, saw both children in their night-dresses, standing near the bed. He immediately jumped out; whereupon they retreated slowly, and with their faces towards him, to the window, where they disappeared. He says, that the circumstance made a great impression upon him at the time; and, indeed, that it was one that could never be effaced; but he did not know what to think of it, not believing in ghosts, and therefore concluded it must have been some extraordinary spectral illusion; especially as his wife heard nothing. It *may* have been so; but that circumstance by no means proves it.

From these varying degrees of susceptibility, or affinity, there seems to arise another consequence, namely, that more than one person may see the same object, and yet see it differently, and I mention this particularly, because it is one of the objections that unreflecting persons make to phenomena of this kind, second sight especially. In the remarkable instance which is recorded to have occurred at Ripley, in the year 1812, to which I shall allude more particularly in a future chapter, much stress was laid on the fact, that the first seer said, "Look at those beasts!" Whilst the second answered, they were "not beasts, but men." In a former chapter, I mentioned the case of a lady, on board a ship, seeing and feeling a sort of blue cloud hanging over her, which afterwards, as it retired, assumed a human form, though still appearing a vapoury substance. Now, possibly, had her receptivity, or the rapport, been greater, she might have seen the distinct image of her dying friend. I have met with several instances of these cloudy figures being seen, as if the spirit had built itself up a form of atmospheric air; and it is remarkable, that when other persons perceived the apparitions that frequented the Seeress of Prevorst, some saw those as cloudy forms, which she saw distinctly attired in the costume they wore when alive; and thus, on some occasions, apparitions are represented as being transparent, whilst on others they have not been distinguishable from the real corporeal body. All these discrepancies, and others, to be hereafter alluded to, are doubtless only absurd to our ignorance; they are the results of physical laws, as absolute, though not so easily ascertained, as those by which the most ordinary phenomena around us are found explicable.

With respect to these cloudy forms, I have met with four instances lately; two occurring to ladies, and two to gentlemen; the one a minister, and the other a man engaged in business; and although I am quite aware that these cases are not easily to be distinguished from those of spectral illusion, yet I do not think them so myself; and as they occurred to persons in their normal state of health, who never before or since experienced anything of the kind, and who could find nothing in their own circumstances to account for its happening then, I shall mention them. In the instances of the gentleman and one of the ladies, they were suddenly awakened, they could not tell by what, and perceived bending over them a cloudy form, which immediately retreated slowly to the other end of the room, and disappeared. In the fourth case, which occurred to an intimate friend of my own, she had not been to sleep; but having been the last person up in the house, had just stept into the bed, where her sister had already been some time asleep. She was perfectly awake, when her attention was attracted by hearing the clink of glass, and on looking up, she saw a figure standing on the hearth, which was exactly opposite her side of the bed, and as there was water and a tumbler there, she concluded that her sister had stept out at the bottom, unperceived by her, and was drinking. Whilst she was carelessly observing the figure, it moved towards the bed, and laid a heavy hand upon her, pressing her arm in a manner that gave her pain. "Oh, Maria, don't!" she exclaimed; but as the form retreated, and she lost sight of it, a strange feeling crept over her, and she stretched out her hand to ascertain if her sister was beside her. She was, and asleep; but this movement awoke her, and she found the other now in considerable agitation. She, of course, tried to persuade her that it was a dream, or night-mare, as did the family the next day; but she was quite clear in her mind at the time, as she then assured me, that it was neither one nor the other; though now, at the distance of a year from the occurrence, she is very desirous of putting that construction upon it. As somebody will be ready to suggest that this was a freak played by one of the family, I can only answer that that is an explanation that no one who is acquainted with all the circumstances, could admit; added to which, the figure did not disappear in the direction of the door, but in quite an opposite one.

A very singular thing happened to the accomplished authoress of "Letters from the Baltic," on which my readers may put what interpretation they please, but I give it here as a pendant to the last story. The night before she left Petersburgh, she passed in the house of a friend. The room appropriated to her use was a large dining-room, in which a temporary bed was placed, and a folding screen was so arranged as to give an air of comfort to the nook where the bed stood. She went to bed, and to sleep, and no one who knows her can suspect her of seeing spectral illusions, or being incapable of distinguishing her own condition when she saw anything whatever. As she was to commence her journey on the following day, she had given orders to be called at an early hour, and, accordingly, she found herself awakened towards morning by an old woman in a complete Russian costume, who looked at her, nodding and smiling, and intimating, as she supposed, that it was time to rise. Feeling, however, very sleepy, and very unwilling to do so, she took her watch from behind her pillow, and, looking at it, perceived that it was only four o'clock. As, from the costume of the old woman, she knew her to be a Russian, and therefore not likely to understand any language she could speak, she shook her head, and pointed to the watch, giving her to understand that it was too early. The woman looked at her, and nodded, and then retreated, whilst the traveller laid down again and soon fell asleep. By and by, she was awakened by a knock at the door, and the voice of the maid whom she had desired to call her. She bade her come in, but the door being locked on the inside, she had to get out of bed to admit her. It now occurred to her to wonder how the old woman had entered, but, taking it for granted there was some other mode of ingress, she did not trouble herself about it, but dressed, and descended to breakfast. Of course, the enquiry usually addressed to a stranger was made, they hoped she had slept well! "Perfectly," she said, "only that one of their good people had been somewhat over anxious to get her up in the morning;" and she then mentioned the old woman's visit, but to her surprise they declared they had no such person in the family. "It must have been some old nurse, or laundress, or somebody of that sort," she suggested. "Impossible!" they answered; "You must have dreamt the whole thing; we have no old woman in the house; nobody wearing that costume; and nobody could have got in, since the door must have been fastened long after that!" And these assertions the servants fully confirmed; added to which, I should observe, the house, like foreign houses in general, consisted of a flat, or floor, shut in by a door, which separated it entirely from the rest of the building, and, being high up from the street, nobody could even have gained access by a window. The lady now beginning to be somewhat puzzled, enquired if there were any second entrance into the room; but, to her surprise, she heard there was not, and she then mentioned that she had locked the door on going to bed, and had found it locked in the morning. The thing has ever remained utterly inexplicable. and the family, who were much more amazed by it than she was, would willingly believe it to have been a dream, but, whatever the interpretation of it may be, she feels quite certain that that is not the true one.

I make no comments on the above case, though a very inexplicable one; and I scarcely know whether to mention any of those well established tales, which appear certainly to be as satisfactorily attested as any circumstance which is usually taken simply on report. I allude, particularly, to the stories of General Wynyard, Lord Tyrone, and Lady Beresford, the case which took place at Havant, in Hampshire, and which is related in a letter from Mr. Caswell, the mathematician, to Dr. Bentley; that which occurred in Cornwall, as narrated by the Rev. Mr. Ruddle, one of the prebendaries of Exeter, whose assistance and advice was asked, and who himself had two interviews with the spirit; and many others, which are already published in different works, especially in a little book entitled "Accredited

Ghost Stories." I may however mention, that with respect to those of Lady Beresford and General Wynyard, the families of the parties have always maintained their entire belief in the circumstances; as do the family of Lady Betty Cobb, who took the ribbon from Lady Beresford's arm, after she was dead; she having always worn it since her interview with the apparition, in order to conceal the mark he had left by touching her.

There have been many attempts to explain away the story of Lord Littleton's warning, although the evidence for it certainly satisfied the family, as we learn from Dr. Johnson, who said, in regard to it, that it was the most extraordinary thing that had happened in his day, and that he heard it from the lips of Lord Westcote, the uncle of Lord Littleton.

There is a sequel however to this story, which is extremely well authenticated, though much less generally known. It appears that Mr. Miles Peter Andrews, the intimate friend of Lord Littleton, was at his house, at Dartford, when Lord L., died at Pitt-place, Epsom, thirty miles off. Mr. Andrews' house was full of company, and he expected Lord Littleton, whom he had left in his usual state of health, to join him the next day, which was Sunday.

Mr. Andrews himself feeling rather indisposed on the Saturday evening, retired early to bed, and requested Mrs. Pigou, one of his guests, to do the honours of his supper-table. He admitted, for he is himself the authority for the story, that he fell into a feverish sleep on going to bed, but was awakened between eleven and twelve by somebody opening his curtains, which proved to be Lord Litttleton, in a night-gown and cap, which Mr. Andrews recognized. Lord L. spoke, saying that he was come to tell him *all was over.* It appears that Lord Littleton was fond of practical joking, and as Mr. A. entertained no doubt whatever of his visitor being Lord L. himself, in the body, he supposed that this was one of his tricks; and, stretching his arm out of bed, he took hold of his slippers, the nearest thing he could get at, and threw them at him, whereupon the figure retreated to a dressing-room, which had no ingress nor egress except through the bed-chamber. Upon this, Mr. Andrews jumped out of bed to follow him, intending to chastise him further, but he could find nobody in either of the rooms, although the door was locked on the inside, so he rang his bell, and enquired who had seen Lord Littleton. Nobody had seen him; but, though how he had got in or out of the room, remained an enigma, Mr. Andrews asserted that he was certainly there; and, angry at the supposed trick, he ordered that they should give him no bed, but let him go and sleep at the inn. Lord Littleton, however, appeared no more, and Mr. Andrews went to sleep, not entertaining the slightest suspicion that he had seen an apparition. It happened that, on the following morning, Mrs. Pigou had occasion to go at an early hour to London, and great was her astonishment to learn that Lord Littleton had died on the preceding night. She immediately dispatched an express to Dartford with the news, upon the receipt of which, Mr. Andrews, then quite well, and remembering perfectly all that had happened, swooned away. He could not understand it, but it had a most serious effect upon him, and, to use his own expression, he was not his own man again for three years. There are various authorities for this story, the correctness of which is vouched for by some members of Mrs. Pigou's family, with whom I am acquainted, who have frequently heard the circumstances detailed by herself, and who assure me it was always believed by the family. I really, therefore, do not see what grounds we have for doubting either of these facts. Lord Westcote, on whose word Dr. Johnson founded his belief of Lord Littleton's warning, was a man of strong sense ; and that the story was not looked upon lightly by the family, is proved by the circumstance that the dowager Lady Littleton had a picture, which was seen by Sir Nathaniel Wraxhall in her house in Portugal-street, as mentioned in his memoirs, wherein the event was commemorated. His Lordship is in bed, the dove appears at the window, and a female figure stands at the foot of the couch, announcing to the unhappy profligate his approaching dissolution. That he mentioned the warning to his valet, and some other persons, and that he talked of *jockeying* the ghost by surviving the time named, is certain; as, also, that he died with his watch in his hand, precisely at the appointed period. Mr. Andrews says, that he was subject to fits of strangulation, from a swelling in the throat, which might have killed him at any moment; but his decease having proceeded from a natural and obvious cause, does not interfere one way or the other with the validity of the prediction, which simply foretold his death at a particular period, not that there was to be anything preternatural in the manner of it.

As I find so many people willing to believe in wraiths, who cannot believe in ghosts—that is, they are overpowered by the numerous examples, and the weight of evidence for the first—it would be very desirable if we could ascertain whether these wraiths are seen before the death occurs, or after it; but, though the day is recorded, and seems always to be the one on which the death took place, and the hour about the same, minutes are not sufficiently observed to enable us to answer that question. It would be an interesting one, because the argument advanced by those who believe that the dead never are seen, is, that it is the strong will and desire of the expiring person which enables him so to act on the nervous system of his distant friend, that the imagination of the latter projects the form, and sees it as if objectively. By *imagination* I do not simply mean to convey the common notion implied by that much abused word, which is only *fancy*, but the *constructive* imagination, which is a much higher function, and which, inasmuch as man is made in the likeness of God, bears a distant relation to that sublime power by which the Creator projects, creates, and upholds his universe; whilst the far-working of the departing spirit seems to consist in the strong will to do, reinforced by the strong faith that the thing can be done. We have rarely the strong will, and still more rarely the strong faith, without which the will remains ineffective. In the following case, which is perfectly authentic, the apparition of Major R. was seen several hours after his death had occurred.

In the year 1785, some cadets were ordered to proceed from Madras, to join their regiments up the country. A considerable part of the journey was to be made in a barge, and they were under the conduct of a senior officer, Major R. In order to relieve the monotony of the voyage, this gentleman proposed, one day, that they should make a shooting excursion inland, and walk round to meet the boat at a point agreed on, which, owing to the windings of the river, it would not reach till evening. They accordingly took their guns, and as they had to cross a swamp, Major R., who was well acquainted with the country, put on a heavy pair of top-boots, which, together with an odd limp he had in his gait, rendered him distinguishable from the rest of the party at a considerable distance. When they reached the jungle, they found there was a wide ditch to leap, which all succeeded in doing except the Major, who being less young and active, jumped short of the requisite distance; and although he scrambled up unhurt, he found his gun so crammed full of wet sand that it would be useless till thoroughly cleansed. He therefore bade them walk on, saying he would follow; and taking off his hat, he sat down in the shade, where they left him. When they had been beating about for game some time, they began to wonder the Major did not come on, and they shouted to let him know where abouts they were; but there was no answer, and hour after hour passed without his appearance, till at length they began to feel somewhat uneasy. Thus the day wore away, and they found themselves approaching the rendezvous; the boat was in sight, and they were walking down to it, wondering how their friend could have missed them, when suddenly, to their great joy, they saw him before them making towards the barge. He was without his hat or gun, limping hastily along, in his top-boots, and did not appear to observe them. They shouted after him, but as he did not look round, they began to run, in order to overtake him; and, indeed, fast as he went, they did gain considerably upon him. Still he reached the boat first, crossing the plank which the boatmen had placed ready for the gentlemen they saw approaching. He ran down the companion stairs, and they after him; but inexpressible was their surprise when they could not find him below. They ascended again, and enquired of the boatmen what had become of him; but they declared he had not come on board, and that nobody had crossed the plank till the young men themselves had done so.

Confounded and amazed at what appeared so inexplicable, and doubly anxious about their friend, they immediately resolved to retrace their steps in search of him; and, accompanied by some Indians who knew the jungle, they made their way back to the spot where they had left him. From thence some footmarks enabled them to trace him, till, at a very short distance from the ditch, they found his hat and his gun. Just then the Indians called out to them to beware, for that there was a sunk well thereabouts, into which they might fall. An apprehension naturally seized them that this might have been the fate of their friend; and on examining its edge, they saw a mark as of a heel slipping up; upon this, one of the Indians consented to go down, having a rope with which they had provided themselves tied round his waist, for, aware of the existence of the wells, the natives suspected what had actually occurred, namely, that the unfortunate gentleman had slipt into one of these traps, which, being overgrown with brambles, were not discernible by the eye. With the assistance of the Indian, the body was brought up and carried back to the boat, amidst the deep regrets of the party, with whom he had been a great favourite. They proceeded with it to the next station, where an enquiry was instituted as to the manner of his death, but of course there was nothing more to be elicited.

I give this story as related by one of the parties present, and there is no doubt of its perfect authenticity. He says, he can in no way account for the mystery—he can only relate the fact; and not one, but the whole *five* cadets, saw him as distinctly as they saw each other. It was evident, from the spot where the body was found, which was not many hundred yards from the well, that the accident must have occurred very shortly after they left him. When the young men reached the boat,

Major R. must have been, for some seven or eight hours, a denizen of the other world, yet he kept the rendezvous!

There was a similar occurrence in Devonshire, some years back, which happened to the well known Dr. Hawker, who, one night, in the street, observed an old woman pass him, to whom he was in the habit of giving a weekly charity. Immediately after she had passed, he felt somebody pull his coat, and, on looking round, saw it was she, whereupon he put his hand in his pocket to seek for a sixpence, but, on turning to give it to her, she was gone. He thought nothing about it, but when he got home, he enquired if she had had her money that week, when, to his amazement, he heard she was dead, but his family had forgotten to mention the circumstance. I have met with two curious cases occurring in Edinburgh, of late years; in one, a young man and his sister were in their kitchen, warming themselves over the fire, before they retired to bed, when, on raising their eyes, they both saw a female figure dressed in white, standing in the doorway, and looking at them; she was leaning against one of the door-posts. Miss E., the young lady, screamed, whereupon the figure advanced, crossed the kitchen towards a closet, and disappeared. There was no egress at the closet; and, as they lived in a flat, and the door was closed for the night, a stranger could neither have entered the house nor got out of it. In the other instance, there were two houses on one flat, the doors opposite each other. In one of the houses there resided a person with her two daughters, grown up women; in the other lived a shoemaker and his wife. The latter died, and it was said her husband had ill-treated her, and worried her out of the world. He was a drunken, dissipated man, and used to be out till a late hour most nights, whilst this goor woman sat up for him; and, when she heard a voice on the stairs, or a bell, she used often to come out and look over, to see if it were her husband returned. One night, when she had been dead some weeks, the two young women were ascending the stairs to their own door, when, to their amazement, they both saw her standing at the top, looking over as she used to do in her life time. At the same moment, their mother opened the door, and saw the figure also; the girls rushed past, overcome with terror, and one, if not both, fainted, as soon as they got into the door. The youngest fell on her face in the passage.

Another case, which occurred in this town, I mention, although I know it is liable to be called a spectral illusion, because it bears a remarkable similarity to one which took place in America. A respectable woman lost her father, for whom she had a great affection; she was of a serious turn, and much attached to the tenets of her church, in which particulars she thought her father had been deficient. She was therefore very unhappy about him, fearing that he had not died in a proper state of mind. A considerable time had elapsed since his death, but her distrust of his condition was still causing her uneasiness, when, one day, whilst she was sitting at her work, she felt something touch her shoulder, and on looking round she perceived her father, who bade her cease to grieve about him, as he was not unhappy. From that moment, she became perfectly resigned and cheerful. The American case—I have omitted to write down the name of the place, and forget it—was that of a mother and son. She was also a highly respectable person, and was described to me as perfectly trustworthy, by one who knew her. She was a widow, and had one son, to whom she was extremely attached, He however disappeared, one day, and she never could learn what had become of him; she always said, that if she did but know his fate, she should be happier. At length, when he had been dead a considerable time, her attention was, one day, whilst reading, attracted by a slight noise, which induced her to look round; and she saw her son, dripping with water, and with a sad expression of countenance. The features however presently relaxed, and they assumed a more pleasing aspect before he disappeared. From that time she ceased to grieve, and it was subsequently ascertained that the young man had run away to sea; but no more was known of him. Certain it was, however, that she attributed her recovered tranquillity to having seen her son as above narrated.

A lady, with whom I am acquainted, when she was a girl, was one day standing at the top of the stairs, with two others, discussing their games, when they each suddenly exclaimed, "Who's that?" There was a fourth among them; a girl in a checked pinafore; but she was gone again. They had all seen her. One day a younger brother, in the same house, was playing with a whip, when he suddenly laughed at something, and cried, "Take that;" and described having seen the same girl. This led to some enquiry, and it was said that such a girl as they described had lived in that house, and had died from the bite of a mad dog; or, rather, had been smothered between two feather beds; but whether that was actually done, or was only a report, I cannot say. Supposing this to have been no illusion, and I really cannot see how it could be one, the memory of past sports and pleasures seems to have so survived, as to have attracted the young soul, prematurely cut off, to the spot where the same sports and pleasures were being enjoyed by the living.

A maid servant, in one of the midland counties of England, being up early one morning, heard her name called in a voice that seemed to be her brother's, a sailor then at sea; and running up, she found him standing in the hall; he said he was come from afar, and was going again, and mentioned some other things, when her mistress, hearing voices, called to know who she was talking to; she said, it was her brother, from sea. After speaking to her for some time, she suddenly lost sight of him, and found herself alone. Amazed and puzzled, she told her mistress what had happened, who being led thus to suspect the kind of visitor it was, looked out of the window to ascertain if there were any marks of footsteps, the ground being covered with snow. There were however none, and it was therefore clear that nobody could have entered the house. Intelligence afterwards arrived of the young man's death.

This last is a case of wraith, but a more complicated one, from the circumstance of speech being superadded. But this is not by any means an isolated particular; there are many such. The author of the book called "Accredited Ghost Stories," whose name I at this moment forget, and I have not the book at hand, gives, on his own authority, the following circumstance, professing to be acquainted with the parties. A company were visiting York Cathedral, when a gentleman and lady, who had detached themselves from the rest, observed an officer wearing a naval uniform approaching them; he walked quickly, saying to the lady as he passed, "There *is* another world." The gentleman, seeing her greatly agitated, pursued the stranger, but lost sight of him, and nobody had seen such a person but themselves. On returning to his companion, she told him that it was her brother, who was then abroad with his ship, and with whom she had frequently held discussions as to whether there was or was not a future life. The news of the young man's death shortly reached the family. In this case, the brother must have been dead; the spirit must have passed out of this world into that other, the existence of which he came to certify. This is one of those cases which, happening not long ago, leads one especially to regret the want of moral courage which prevents people giving up their names, and avowing their experience. The author of the above-mentioned book, from which I borrow this story, says, that the sheet had gone to the press with the real names of the parties attached, but that he was requested to withdraw them, as it would be painful to the family. My view of this case is so different, that, had it occurred to myself, I should have felt it my imperative duty to make it known, and give every satisfaction to enquirers.

Some years ago, during the war, when Sir Robert H. E. was in the Netherlands, he happened to be quartered with two other officers, one of whom was dispatched into Holland on an expedition. One night, during his absence, Sir R. H. E. awoke, and, to his great surprise, saw this absent friend sitting on the bed, which he used to occupy, with a wound in his breast. Sir R. immediately awoke his companion, who saw the spectre also. The latter then addressed them, saying, that he had been that day killed in a skirmish, and that he had died in great anxiety about his family, wherefore he had come to communicate that there was a deed of much consequence to them deposited in the hands of a certain lawyer in London, whose name and address he mentioned, adding that this man's honesty was not to be altogether relied on. He therefore requested that, on their return to England, they would go to his house and demand the deed, but that, if he denied the possession of it, they were to seek it in a certain drawer in his office, which he described to them. The circumstance impressed them very much at the time, but a long while had elapsed ere they reached England, during which period they had gone through so many adventures and seen so many friends fall around them, that this impression was considerably weakened, insomuch that each went to his own home and his own pursuits without thinking of fulfilling the commission they had undertaken. Some time afterwards, however, it happened that they both met in London, and they then resolved to seek the street that had been named to them, and ascertain if such a man lived there. They found him, requested an interview, and demanded the deed, the possession of which he denied; but their eyes were upon the drawer that had been described to them; where they asserted it to be; and being there discovered, it was delivered into their hands. Here, also, the soul had parted from the body, whilst the memory of the past and an anxiety for the worldly prosperity of those left behind, survived; and we thus see that the condition of mind in which this person had died, remained unchanged. He wasnot indifferent to the worldly prosperity of his relatives, and he found his own state rendered unhappy by the fear that they might suffer from the dishonesty of his agent. It may here be naturally objected that hundreds of much loved widows and orphans have been ruined by dishonest trustees and agents, where no ghost came back to instruct them in the means of obviating the misfortune. This is, no doubt, a very legitimate objection, and one which it is very difficult to answer. I must, however, repeat what I said before; nature is full of exceptional cases, whilst we know very little of the laws which regulate these exceptions; but we may see a very good reason for the fact that such communications *are* the exception, and not the rule; for if they were the latter the whole economy of this earthly life would be overturned, and its affairs must necessarily be conducted in a totally different manner to that which prevails at present. What the effects of such an arrangement of nature would be, had it pleased God to make it, he alone knows; but certain it is, that man's freedom, as a moral agent, would be in a great degree abrogated, were the barriers that impede our intercourse with the spiritual world removed.

It may be answered, that this is an argument which may be directed against the fact of such appearances being permitted, at all; but that is a fallacious objection. Earthquakes and hurricanes are occasionally permitted, which overthrow the work of man's hands for centuries; but if these convulsions of nature were of every day occurrence, nobody would think it worth their while to build a house or cultivate the earth, and the world would be a wreck and a wilderness. The apparitions that do appear, are not without their use to those who believe in them; whilst there is too great an uncertainty attending the subject, generally, to allow of its ever being taken into consideration in mundane affairs.

The old, so called, superstition of the people, that a person's "dying with something on his mind," is one of the frequent causes of these revisitings, seems, like most other of their superstitions, to be founded on experience. I meet with many cases in which some apparently trivial anxiety, or some frustrated communication, prevents the uneasy spirit flinging off the bonds that bind it to the earth. I could quote many examples characterised by this feature, but will confine myself to two or three.

Jung Stilling gives a very curious one, which occurred in the year 1746, and for the authenticity of which he vouches. A gentleman, of the name of Dorrien, of most excellent character and amiable disposition, who was tutor in the Carolina Colleges, at Brunswick, died there in that year; and, immediately previous to his death, he sent to request an interview with another tutor, of the name of Hofer, with whom he had lived on terms of friendship. Hofer obeyed the summons, but came too late; the dying man was already in the last agonies. After a short time, rumours began to circulate that Herr Dorrien had been seen by different persons about the college; but as it was with the pupils that these rumours originated, they were supposed to be mere fancies, and no attention whatever was paid to them. At length, however, in the month of October, three months after the decease of Herr Dorrien, a circumstance occurred that excited considerable amazement amongst the professors. It formed part of the duty of Hofer to go through the college every night, between the hours of eleven and twelve, for the purpose of ascertaining that all the scholars were in bed, and that nothing irregular was going on amongst them. On the night in question, on entering one of the anti-rooms in the execution of this duty, he saw, to his great amazement, Herr Dorrien, seated, in the dressing-gown and white cap he was accustomed to wear, and holding the latter with his right hand, in such a manner as to conceal the upper part of the face; from the eyes to the chin, however, it was distinctly visible. This unexpected sight naturally startled Hofer, but, summoning resolution, he advanced into the young men's chamber, and, having ascertained that all was in order, closed the door; he then turned his eyes again towards the spectre, and there it sat as before, whereupon he went up to it, and stretched out his arm towards it; but he was now seized with such a feeling of indescribable horror, that he could scarcely withdraw his hand, which became swollen to a degree that for some months he had no use of it. On the following day, he related this circumstance to the professor of mathematics, Oeder, who of course treated the thing as a spectral illusion. He, however, consented to accompany Hofer on his rounds the ensuing night, satisfied that he should be able either to convince him it was a mere phantasm, or else a spectre of flesh and blood who was playing him a trick. They accordingly went at the usual hour, but no sooner had the professor of mathematics set his foot in that same room, than he exclaimed, "By Heavens, it is Dorrien himself!" Hofer, in the mean time, proceeded into the chamber as before, in the pursuance of his duties, and, on his return, they both contemplated the figure for some time; they had, however, neither of them the courage to address or approach it, and finally quitted the room, very much impressed, and perfectly convinced that they had seen Dorrien. This incident soon got spread abroad, and many people came in hopes of satisfying their own eyes of the fact, but their pains were fruitless; and even Professor Oeder, who had made up his mind to speak to the apparition, sought it repeatedly in the same place in vain. At length, he gave it up, and ceased to think of it, saying, "I have sought the ghost long enough; if he has anything to say, he must now seek me." About a fortnight after this, he was suddenly awakened, between three and four o'clock in the morning, by something moving in his chamber, and, on opening his eyes, he beheld a shadowy form, having the same appearance as the spectre, standing in front of a press which was not more than two steps from his bed. He raised himself, and contemplated the figure, the features of which he saw distinctly for some minutes, till it disappeared. On the following night he was awakened in the same manner, and saw the figure as before, with the addition that there was a sound proceeded from the door of the press, as if somebody was leaning against it. The spectre also staid longer this time, and Professor Oeder, no doubt frightened and angry, addressing it as an evil spirit, bade it begone, whereon it made gestures with its head and hands that alarmed him so much, that he adjured it, in the name of God, to leave him, which it did. Eight days now elapsed without any further disturbance, but, after that period, the visits of the spirit were resumed, and he was awakened by it repeatedly about three in the morning, when it would advance from the press to the bed, and, hang its head over him in a manner so annoying, that he started up and struck at it, whereupon it would retire, but presently advance again. Perceiving, now, that the countenance was rather placid and friendly than otherwise, the professor at length addressed it, and, having reason to believe that Dorrien had left some debts unpaid; he asked him if that were the case, upon which the spectre retreated some steps, and seemed to place itself in an attitude of attention. Oeder reiterated the enquiry, whereupon the figure drew its hand across its mouth, in which the professor now observed a short pipe. "Is it to the barber you are in debt?" he enquired. The spectre slowly shook its head. "Is it to the tobacconist, then?" asked he, the question being suggested by the pipe. Hereupon the form retreated, and disappeared. On the following day, Oeder narrated what had occurred to Councillor Erath, one of the curators of the college, and also to the sister of the deceased, and arrangements were made for discharging the debt. Professor Seidler, of the same college, now proposed to pass the night with Oeder for the purpose of observing if the ghost came again, which it did about five o'clock, and awoke Oeder as usual, who awoke his companion, but just then the form disappeared, and Seidler said he only saw something white. They then both disposed themselves to sleep, but presently Seidler was aroused by Oeder's starting up and striking out, whilst he cried, with a voice expressive of rage and horror, "Begone! You have tormented me long enough! If you want anything of me, say what it is, or give me an intelligible sign, and come heie no more!"

Seidler heard all this, though he saw nothing; but as soon as Oeder was somewhat appeased, he told him that the figure had returned, and not only approached the bed, but stretched itself upon it. After this, Oeder burnt a light, and had some one in the room with him every night. He gained this advantage by the light, that he saw nothing; but between the hours of three and five, he was generally awakened by noises in his room, and other symptoms that satisfied him the ghost was there. At length, however, this annoyance ceased also; and trusting that his unwelcome guest had taken his leave, he dismissed his bedfellow, and dispensed with his light. Two nights passed quietly over: on the third, however, the spectre returned; but very perceptibly darker. It now presented another sign, or symbol, which seemed to represent a picture, with a hole in the middle, through which it thrust its head. Oeder was now so little alarmed, that he bade it express its wishes more clearly, or approach nearer. To these requisitions the apparition shook its head, and then vanished. This strange phenomenon recurred several times, and even in the presence of another curator of the college; but it was with considerable difficulty they discovered what the symbol was meant to convey. They at length, however, found that Dorrien, just before his illness, had obtained, on trial, several pictures for a magic lantern, which had never been returned to their owner. This was now done, and from that time the apparition was neither seen nor heard again. Professor Oeder made no secret of these circumstances; he related them publicly in court and college; he wrote the account to several eminent persons, and declared himself ready to attest the facts upon his oath.

Stilling, who relates this story, has been called superstitious; he may be so; but his piety and his honesty are above suspicion; he says the facts are well known, and that he can vouch for their authenticity; and as he must have been a cotemporary of the parties concerned, he had, doubtless, good opportunities of ascertaining what foundation there was for the story. It is certainly a very extraordinary one, and the demeanour of the spirit as little like what we should have naturally apprehended as possible; but, as I have said before, we have no right to pronounce any opinion on this subject, except from experience, and there are two arguments to be advanced in favour of this narration; the one being, that I cannot imagine anybody setting about to invent a ghost-story, would have introduced circumstances so apparently improbable and inappropriate; and the other consisting in the fact, that I have met with numerous relations, coming from very opposite quarters, which seem to corroborate the one in question.

With respect to the cause of the spectre's appearance, Jung Stilling, I think, reasonably enough, suggests, that the poor man had intended to commission Hofer to settle these little affairs for him, but that delaying this duty too long, his mind had been oppressed by the recollection of them in his last moments—he had carried his care with him, and it bound him to the earth. Wherefore, considering how many persons die with duties unperformed, this anxiety to repair the neglect, is not more frequently manifested, we do not know; some reasons we have already suggested as possible; there may be others of which we can form no idea, any more than we can solve the question, why in some cases communication and even speech seems easy, whilst in this instance, the spirit was only able to convey its wishes by gestures and symbols. Its addressing itself to Oeder instead of Hofer, probably arose from its finding communication with him less difficult; the swelling of Hofer's arm indicating that his physical nature, was not adapted for this spiritual intercourse. With respect to Oeder's expedient of burning a light in his room, in order to prevent his seeing this shadowy form, we can comprehend, that the figure would be discerned more easily on the dark ground of comparative obscurity, and that clear light would render it invisible. Dr. Kerner mentions, on one occasion, that whilst sitting in an adjoining room, with the door open, he had seen a shadowy figure, to whom his patient was speaking, standing beside her bed; and catching up a candle, he had rushed towards it; but as soon as he had thus illuminated the chamber, he could no longer distinguish it.

The ineffective and awkward attempts of this apparition, to make itself understood, are not easily to be reconciled to our ideas of a spirit, whilst at the same time, that which it could do, and that which it could not—the powers it possessed and those it wanted—tend to throw some light on its condition. As regards space, we may suppose, that in this instance, what St. Martin said of ghosts in general, may be applicable, "*Je ne crois pas aux revenants, mais je croix aux restants;*" that is, he did not believe, that spirits who had once quitted the earth, returned to it, but he believed that some did not quit it, and thus, as the somnambule mentioned in a former chapter said to me, "Some are waiting and some are gone on before." Dorrien's uneasiness and worldly care chained him to the earth, and he was a *restant*, but, being a spirit, he was inevitably inducted into some of the inherent properties of spirit; matter to him was no impediment, neither doors nor walls could keep him out; he had the intuitive perception of whom he could most easily communicate with, or he was brought into rapport with Oeder by the latter's seeking him; and he could either so act on Oeder's constructive imagination, as to enable it to project his own figure, with the short pipe and the pictures, or he could, by the magical power of his will, build up these images out of the constituents of the atmosphere. The last seems the most probable, because, had the rapport with Oeder, or Oeder's receptivity, been sufficient to enable the spirit to act potently upon him, it would have been also able to infuse into his mind the wishes it desired to convey, even without speech, for speech, as a means of communication betwixt spirits, must be quite unnecessary. Even in spite of these dense bodies of ours, we have great difficulty in concealing our thoughts from each other; and the somnambule reads the thoughts not only of his magnetiser, but of others, with whom he is placed in rapport. In cases where speech appears to be used by a spirit, it is frequently not audible speech, but only this transference of thought, which appears to be speech from the manner in which the thought is borne in and enters the mind of the receiver; but it is not through his ears, but through his universal supplementary sense, that he receives it; and it is no more like what we mean by *hearing*, than is the seeing of a *clairvoyant*, or a spirit, like our seeing by means of our bodily organs. In those cases where the speech is audible to other persons, we must suppose that the magical will of the spirit can, by means of the atmosphere, simulate these sounds as it can simulate others, of which I shall have to treat by and by. It is remarkable, that, in some instances, this magical power seems to extend so far as to represent to the eye of the seer a form apparently so real, solid, and life-like, that it is not recognizable from the living man; whilst in other cases the production of a shadowy figure seems to be the limit of its agency, whether limited by its own faculty, or the receptivity of its subject; but we must be quite sure that the form is, in either instance, equally ethereal or immaterial. And it will not be out of place here to refer to the standing joke of the sceptics, about ghosts appearing in coats and waistcoats. Bentham thought he had settled the question for ever by that objection; and I have heard it since frequently advanced by very acute persons, but, properly considered, it has not the least validity.

Whether or not the soul on leaving its earthly tabernacle finds itself at once clothed with that spiritual body, which St. Paul refers to, is what we cannot know, though it seems highly probable; but if it be so, we must be sure that this body resembles in its nature, that fluent subtle kind of matter, called by us imponderables, which are capable of penetrating all substances; and unless there be no visible body at all, but only the will of a disembodied spirit acting upon one yet in the flesh, in which case it were as easy to impress the imagination with a clothed figure as an unclothed one, we must conclude that this ethereal flexible form, whether permanent or temporary, may be held together and retain its shape by the volition of the spirit, as our bodies are held together by the principle of life that is in them; and we see in various instances, where the spectator has been bold enough to try the experiment, that though the shadowy body was pervious to any substance passed through it, its integrity was only momentarily interrupted, and it immediately recovered its previous shape. Now, as a spirit, provided there be no especial law to the contrary, partial or universal, absolute or otherwise, governing the spiritual world—must be where its thoughts and wishes are, just as we should be at the place we intently think of, or desire, if our solid bodies did not impede us, so must a spirit appear as it is, or as it *conceives* of itself; morally, it can only conceive of itself as it is, good or bad, light or dark; but it may conceive of itself clothed as well as unclothed; and if it can conceive of its former body, it can equally conceive of its former habiliments; and so represent them, by its power of will to the eye, or present them to the constructive imagination of the seer; and it will be able to do this with a degree of distinctness proportioned to the receptivity of the latter, or to the intensity of the rapport which exists between them. Now, considered in this way, the appearance of a spirit "in its habit as it lived," is no more extraordinary than the appearance of a spirit at all, and it adds no complexity to the phenomenon. If it appears at all, in a recognizable form, it must come naked or clothed; the former, to say the least of it, would be much more frightful and shocking; and if it be clothed, I do not see what right we have to expect it shall be in a fancy costume, conformable to our ideas, which are no ideas at all, of the other world; nor why, if it be endowed with the memory of the past, it should not be natural to suppose it would assume the external aspect it wore, during its earthly pilgrimage. Certain it is, whether consistent with our notions or not, all tradition seems to show that this is the appearance they assume; and the very fact, that on the first view of the case, and until the question is philosophically considered, the addition of a suit of clothes to the phenomenon, not only renders its acceptance much more difficult, but throws an air of absurdity and improbability on the whole subject, furnishes a very strong argument in favour of the persuasion, that this notion has been founded on experience, and is not the result either of fancy or gratuitous invention. The idea of spirits appearing like angels, with wings, &c. seems to be drawn from these relations in the Bible, when messengers were sent from God to man; but those departed spirits are not angels, though probably destined in the course of ages to become so; in the mean time, their moral state continues as when they quitted the body, and their memories and affections are with the earth, and so, earthly they appear, more or less. We meet with some instances in which bright spirits have been seen; protecting spirits, for example, who have shaken off their earth entirely, clinging to it yet but by some holy affection or mission of mercy, and these appear, not with wings, which whenever seen are merely symbolical, for we cannot imagine they are necessary to the motion of a spirit, but clothed in robes of light. Such appearances, however, seem much more rare than the others. It will seem to many persons very inconsistent with their ideas of the dignity of a spirit that they should appear and act in the manner I have described, and shall describe further; and I have heard it objected that we cannot suppose God would permit the dead to return merely to frighten the living, and that it is showing him little reverence to imagine he would suffer them to come on such trifling errands, or demean themselves in so undignified a fashion. But God permits men of all degrees of wickedness, and of every kind of absurdity, to exist, and to harrass and disturb the earth, whilst they expose themselves to its obloquy or its ridicule.

Now, as I have observed in a former chapter, there is nothing more perplexing to us in regarding man as a responsible being, than the degree to which we have reason to believe his moral nature is influenced by his physical organization; but leaving this difficult question to be decided—if ever it can be decided in this world—by wiser heads than mine, there is one thing of which we may rest perfectly assured, namely, that let the fault of an impure, or vicious, or even merely sensuous life, lie where it will—whether it be the wicked spirit within, or the ill-organized body without, or a *tertium quid* of both combined, still, the soul that has been a party to this earthly career, must be soiled and deteriorated by its familiarity with evil; and there seems much reason to believe that the dissolution of the connexion between the soul and body, produces far less change in the former than has been commonly supposed. People generally think, if they think on the subject at all, that as soon as they are dead, provided they have lived tolerably virtuous lives, or indeed been free from any great crimes, they will immediately find themselves provided with wings, and straightway fly up to some delightful place, which they call heaven, forgetting how unfit they are for heavenly fellowship; and although I cannot help thinking that the Almighty has mercifully permitted occasional relaxations of the boundaries that separate the dead from the living, for the purpose of showing us our error, we are determined not to avail ourselves of the advantage. I do not mean that these spirits—these *revenants* or *restants*—are special messengers sent to warn us; I only mean that their occasionally "revisiting the glimpses of the moon" form the exceptional cases in a great general law of nature, which divides the spiritual from the material world; and that in framing this law, these exceptions may have been designed for our benefit.

There are several stories extant in the English, and a vast number in the German records, which, supposing them to be well founded—and I repeat, that for many of them we have just as good evidence as for anything else we believe as hearsay or tradition—would go to confirm the fact that the spirits of the dead are sometimes disturbed by what appear to us very trifling cares. I give the following case from Dr. Kerner, who says it was related to him by a very respectable man, on whose word he can entirely rely.

"I was," said Mr. St. S., of S—, "the son of a man who had no fortune but his business, in which he was ultimately successful. At first, however, his means being narrow, he was perhaps too anxious and inclined to parsimony; so that when my mother, careful housewife as she was, asked him for money, the demand generally led to a quarrel. This occasioned her great uneasiness, and having mentioned this characteristic of her husband to her father, the old man advised her to get a second key made to the money-chest, unknown to her husband, considering this expedient allowable and even preferable to the destruction of their conjugal felicity, and feeling satisfied that she would make no ill use of the power possessed. My mother followed his advice, very much to the advantage of all parties; and nobody suspected the existence of this second key, except myself, whom she had admitted into her confidence. Two and twenty years my parents lived happily together, when I, being at the time about eighteen hours journey from home, received a letter from my father informing me that she was ill; that he hoped for her speedy amendment, but that if she grew worse he would send a horse to fetch me home to see her. I was extremely busy at that time, and therefore waited for further intelligence, and as several days elapsed without any reaching me, I trusted my mother was convalescent. One night, feeling myself unwell, I had lain down on the bed with my clothes on to take a little rest. It was between eleven and twelve o clock, and I had not been to sleep, when some one knocked at the door, and my mother entered, dressed as she usually was. She saluted me, and said, 'We shall see each other no more in this world, but I have an injunction to give you. I have given that key to R. (naming a servant we then had), and she will remit it to you. Keep it carefully, or throw it into the water, but never let your father see it; it would trouble him. Farewell, and walk virtuously through life!' And with these words she turned and quitted the room by the door, as she had entered it. I immediately arose, called up my people, expressed my apprehension that my mother was dead, and, without further delay, started for home. As I approached the house, R., the maid, came out, and informed me that my mother had expired betwixt the hours of eleven and twelve on the preceding night. As there was another person present at the moment, she said nothing further to me, but she took an early opportunity of remitting me the key, saying that my mother had given it to her just before she expired, desiring her to place it in my hands, with an injunction that I should keep it carefully, or fling it into the water, so that my father might never know anything about it. I took the key, kept it for some years, and at length threw it into the Lahne."

I am aware that it may be objected by those who believe in wraiths, but in no other kind of apparition, that this phenomenon occurred before the death of the lady, and that it was produced by her energetic anxiety with regard to the key; it may be so, or it may not; but at all events, we see in this case how a comparatively trifling uneasiness may disturb a dying person, and how therefore if memory remains to them, they may carry it with them, and seek by such means as they have, to obtain relief from it.

A remarkable instance of anxiety for the welfare of those left behind, is exhibited in the following story, which I received fron a member of the family concerned:—Mrs. R., a lady very well connected, lost her husband when in the prime of life, and found herself with fourteen children, unprovided for. The overwhelming nature of the calamity depressed her energies to such a degree as to render her incapable of those exertions which could alone redeem them from ruin. The flood of misfortune seemed too strong for her, and she yielded to it without resistance. She had thus given way to despondency some time, when one day, as she was sitting alone, the door opened and her mother, who had been a considerable time dead, entered the room and addressed her, reproving her for this weak indulgence of useless sorrow, and bidding her exert herself for the sake of her children. From that period she threw off the depression, set actively to work to promote the fortunes of her family, and succeeded so well that they ultimately emerged from all their difficulties. I asked the gentleman who related this circumstance to me, whether he believed it. He answered that he could only assure me that she herself affirmed the fact, and that she avowedly attributed the sudden change in her character and conduct to this cause—for his own part, he did not know what to say—finding it difficult to believe in the possibility of such a visit from the dead.

A somewhat similar instance is related by Dr. Kerner, which, he says, he received from the party himself, a man of sense and probity. This gentleman, Mr. F., at an early age lost his mother. Two and twenty years afterwards he formed an attachment to a young person, whose hand he resolved to ask in marriage. Having, one evening, seated himself at his desk, for the purpose of writing his proposal, he was amazed, on accidentally lifting his eyes from the paper, to see his mother looking exactly as if alive, seated opposite to him; whilst she, raising her finger with a warning gesture, said, "Do not that thing!" Not the least alarmed, Mr. F. started up to approach her, whereupon she disappeared. Being very much attached to the lady however, he did not feel disposed to follow her counsel; but having read the letter to his father, who highly approved of the match and who laughed at the ghost, he returned to his chamber to seal it, when whilst he was adding the superscription, she again appeared as before, and reiterated her injunction. But love conquered; the letter was dispatched, the marriage ensued, and after ten years of strife and unhappiness was dissolved by a judicial process.

A remarkable circumstance occurred, about forty years ago, in the family of Dr. Paulus at Stuttgard. The wife of the head of the family having died, they, with some of their connexions, were sitting at table a few days afterwards, in the room adjoining that in which the corpse lay, when, suddenly the door of the latter apartment opened, and the figure of the mother, clad in white robes, entered, and saluting them as she passed, walked slowly and noiselessly through the room, and then disappeared again through the door by which she had entered. The whole company saw the apparition; but the father who was at that time quite in health, died eight days afterwards.

Madame R. had promised an old woodcutter, who had a particular horror of dying in the poor-house, because he knew his body would be given to the surgeons, that she would take care to see him properly interred. The old man lived some years afterwards, and she had quite lost sight of him, and indeed forgotten the circumstance, when she was one night awakened by the sound of some one cutting wood in her bed-chamber; and so perfect was the imitation, that she heard every log flung aside as separated. She started up, exclaiming, "The old man must be dead!" and so it proved; his last anxiety having been that Madame R. should remember her promise.

That our interest in whatever has much concerned us in this life, accompanies us beyond the grave, seems to be proved by many stories I meet with, and the following is of undoubted authenticity:—Some years ago, a music-master died at Erfrert at the age of seventy. He was a miser, and had never looked with very friendly eyes on Professor Rinck, the composer who he knew was likely to succeed to his classes. The old man had lived and died in an apartment adjoining the class-room; and the first day that Rinck entered on his office, whilst the scholars were singing *Aus der tiefe ruf ich dich*, which is a paraphrase of the *De profundis*, he thought he saw through a hole or bull's eye there was in the door something moving about the inner chamber. As the room was void of every kind of furniture, and nobody could possibly be in it, Rinck looked more fixedly; when he distinctly saw a shadow, whose movements were accompanied by a strange rustling sound. Perplexed at the circumstance, he told his pupils that on the following day he should require them to repeat the same choral. They did so; and whilst they were singing, Rinck saw a person walking backwards and forwards in the next room, who frequently approached the hole in the door. Very much struck with so extraordinary a circumstance, Rinck had the choral repeated on the ensuing day; and this time his suspicions were fully confirmed; the old man, his predecessor, approaching the door, and gazing steadfastly into the class-room. "His face," said Rinck, in relating the story to Dr. Mainzer, who has obligingly furnished it to me as entered in his journal at the time, "his face was of an ashy grey. The apparition," he added "never more appeared to me, although I frequently had the choral repeated."

"I am no believer in ghost-stories," he added,

"nor in the least superstitious; nevertheless I cannot help admitting that I have seen this, it is impossible for me ever to doubt or to deny that which I know I saw."

CHAPTER X.

THE FUTURE THAT AWAITS US.

In all ages of the world, and in all parts of it, mankind have earnestly desired to learn the fate that awaited them when they had "shuffled off this mortal coil;" and those pretending to be their instructors have built up different systems which have stood in the stead of knowledge, and more or less satisfied the bulk of the people. The interest on this subject is, at the present period, in the most highly civilized portions of the globe, less than it has been at any preceding one. The great proportion of us live for this world alone, and think very little of the next; we are in too great a hurry of pleasure or business to bestow any time on a subject of which we have such vague notions—notions so vague, that, in short, we can scarcely by any effort of the imagination bring the idea home to ourselves; and when we are about to die we are seldom in a situation to do more than resign ourselves to what is inevitable, and blindly meet our fate; whilst, on the other hand, what is generally called the religious world, is so engrossed by its struggles for power and money, or by its sectarian disputes and enmities; and so narrowed and circumscribed by dogmatic orthodoxies, that it has neither inclination nor liberty to turn back or look around, and endeavour to gather up from past records and present observation such hints as are now and again dropt in our path, to give us an intimation of what the truth may be. The rationalistic age, too, out of which we are only just emerging, and which succeeded one of gross superstition, having settled, beyond appeal, that there never was such a thing as a ghost—that the dead never do come back to tell us the secrets of their prison-house, and that nobody believes such idle tales but children and old women, seemed to have shut the door against the only channel through which any information could be sought. Revelation tells us very little on this subject, reason can tell us nothing; and if nature is equally silent, or if we are to be deterred from questioning her from the fear of ridicule, there is certainly no resource left for us but to rest contented in our ignorance; and each wait till the awful secret is disclosed to ourselves. A great many things have been pronounced untrue and absurd, and even impossible, by the highest authorities in the age in which they lived, which have afterwards, and indeed within a very short period, been found to be both possible and true. I confess myself, for one, to have no respect whatever for these dogmatic denials and affirmations, and I am quite of opinion that vulgar incredulity is a much more contemptible thing than vulgar credulity. We know very little of what *is*, and still less of what may be; and till a thing has been proved, by induction logically impossible, we have no right whatever to pronounce that it is so. As I have said before, *à priori* conclusions are perfectly worthless; and the sort of investigation that is bestowed upon subjects of the class of which I am treating, something worse; inasmuch as they deceive the timid and the ignorant, and that very numerous class which pins its faith on authority and never ventures to think for itself, by an assumption of wisdom and knowledge, which, if examined and analysed, would very frequently prove to be nothing more respectable than obstinate prejudice and rash assertion.

For my own part, I repeat, I insist upon nothing. The opinions I have formed from the evidence collected, may be quite erroneous; if so, as I seek only the truth, I shall be glad to be undeceived and shall be quite ready to accept a better explanation of these facts, whenever it is offered to me; but it is in vain to tell me that this explanation is to be found in what is called imagination, or in a morbid state of the nerves, or an unusual excitement of the organs of colour and form, or in imposture; or in all these together. The existence of all such sources of error and delusion, I am far from denying, but I find instances that it is quite impossible to reduce under any one of those categories, as we at present understand them. The multiplicity of these instances, too—for not to mention the large number that are never made known or carefully concealed, if I were to avail myself liberally of cases already recorded in various works, many of which I know, and many others I hear of as existing, but which I cannot conveniently get access to, I might fill volumes—German literature abounds in them—the number of the examples, I repeat, even on the supposition that they are not facts, would of itself form the subject of a very curious physiological or psychological enquiry. If so many people in respectable situations of life, and in apparently a normal state of health, are either capable of such gross impostures, or the subjects of such extraordinary spectral illusions, it would certainly be extremely satisfactory to learn something of the conditions that induce these phenomena in such abundance; and all I expect from my book at present is, to induce a suspicion that we are not quite so wise as we think ourselves; and that it might be worth while to enquire a little seriously into reports, which may perchance turn out to have a deeper interest for us, than all those various questions, public and private, put together, with which we are d ily agit ting ourselves.

I have alluded in an earlier partaof this work, to the belief entertained by the ancients, that the souls of men on being disengaged from the bodies, passed into a middle state, called Hades, in which their portions seemed neither to be that of complete happiness nor of insupportable misery. They retained their personality, their human form, their memory of the past, and their interest in those that had been dear to them on earth. Communications were occasionally made by the dead to the living; they mourned over their duties neglected and their errors committed; many of their mortal feelings, passions and propensities, seemed to survive; and they sometimes sought to repair, through the instrumentality of the living, the injuries they had formerly inflicted. In short, death was merely a transition from one condition of life to another; but in this latter state, although we do not see them condemned to undergo any torments, we perceive that they are not happy. There are indeed compartments in this dark region; there is Tartarus for the wicked, and the Elysian fields for the good, but they are comparatively thinly peopled. It is in the mid region that these pale shades abound, consistently with the fact, that here on earth, moral, as well as intellectual, mediocrity is the rule; and extremes of good or evil the exceptions.

With regard to the opinion entertained of a future state by the Hebrews, the Old Testament gives us very little information; but what glimpses we do obtain of it, appears to exhibit notions analogous to those of the heathen nations, inasmuch as that the personality and the form seem to be retained, and the possibility of these departed spirits revisiting the earth and holding commune with the living is admitted. The request of the rich man, also, that Lazarus might be sent to warn his brethren, yet alive, of his own miserable condition, testifies to the existence of these opinions; and it is worthy of remark, that the favour is denied, not because its performance is impossible, but because the mission would be unavailing—a prediction which, it appears to me, time has singularly justified. Altogether, the notion that in the state entered upon after we leave this world, the personality and form are retained, that these shades sometimes revisit the earth, and that the memory of the past still survives, seems to be universal; for it is found to exist amongst all people, savage and civilized; and if not founded on observation and experience, it becomes difficult to account for such unanimity on a subject which I think, speculatively considered, would not have been productive of such results; and one proof of this is, that those who reject such testimony and tradition as we have in regard to it, and rely only on their own understandings, appear to be pretty uniformly led to form opposite conclusions. They cannot discern the mode of such a phenomenon; it is open to all sorts of scientific objections, and the *cui bono* sticks in their teeth.

This position being admitted, as I think it must be, we have but one resource left, whereby to account for the universability of this persuasion; which is, that in all periods and places, both mankind and womenkind, as well in health as in sickness, have been liable to a series of spectral illusions of a most extraordinary and complicated nature, and bearing such a remarkable similarity to each other, in regard to the objects supposed to be seen or heard, that they have been universally led to the same erroneous interpretation of the phenomenon. It is manifestly not impossible that this may be the case; and if it be so, it becomes the business of physiologists to enquire into the matter, and give us some account of it. In the mean time, we may be permitted to take the other view of the question, and examine what probabilities seem to be in its favour.

When the body is about to die, that which

cannot die, and which, to spare words, I will call *the soul*, departs from it; whither? We do not know; but, in the first place, we have no reason to believe that the space destined for its habitation is far removed from the earth, since, knowing nothing about it, we are equally entitled to suppose the contrary; and, in the next, that which we call distance is a condition that merely regards material objects, and of which a spirit is quite independent, just as our thoughts are, which can travel from here to China, and back again, in a second of time. Well, then, supposing this being to exist somewhere, and it is not unreasonable to suppose that the souls of the inhabitants of each planet continue to hover within the sphere of that planet, to which, for anything we can tell, they may be attached by a magnetic attraction, supposing it to find itself in space, free of the body, endowed with the memory of the past, and consequently with a consciousness of its own deserts, able to perceive that which we do not ordinarily perceive, namely, those who have passed into a similar state with itself, will it not naturally seek its place amongst those spirits which most resemble itself, and with whom, therefore, it must have the most affinity? On earth, the good seek the good, and the wicked the wicked: and the axiom that "like associates with like," we cannot doubt will be as true hereafter as now. "In my father's house there are many mansions," and our intuitive sense of what is fit and just must needs assure us that this is so. There are too many degrees of moral worth and of moral unworth amongst mankind, to permit of our supposing that justice could be satisfied by an abrupt division into two opposite classes. On the contrary, there must be infinite shades of desert, and, as we must consider that that which a spirit enters into on leaving the body, is not so much a *place* as a *condition*, so there must be as many degrees of happiness or suffering as there are individuals, each carrying with him his own Heaven or Hell. For it is a vulgar notion to imagine that Heaven and Hell are *places;* they are states; and it is in ourselves we must look for both. When we leave the body, we carry them with us; "as the tree falls, so it shall lie." The soul which here has wallowed in wickedness or been sunk in sensuality, will not be suddenly purified by the death of the body; its moral condition remains what its earthly sojourn has trained it to, but its means of indulging its propensities are lost. If it has had no godly aspirations here, it will not be drawn to God there; and if it has so bound itself to the body that it has known no happiness but that to which the body ministered, it will be incapable of happiness when deprived of that means of enjoyment. Here we see at once what a variety of conditions must necessarily ensue; how many comparatively negative states there must be betwixt those of positive happiness or positive misery.

We may thus conceive how a soul, on entering upon this new condition, must find its own place or state; if its thoughts and aspirations here have been heavenward, and its pursuits noble, its conditions will be heavenly. The contemplation of God's works, seen not as by our mortal eyes, but in their beauty and their truth, and ever-glowing sentiments of love and gratitude, and, for aught we know, good offices to souls in need, would constitute a suitable heaven, or happiness for such a being; an incapacity for such pleasures, and the absence of all others, would constitute a negative state, in which the chief suffering would consist in mournful regrets and a vague longing for something better, which the untrained soul that never lifted itself from the earth, knows not how to seek; whilst malignant passions and unquenchable desires would constitute the appropriate hell of the wicked; for we must remember, that although a spirit is independent of those physical laws which are the conditions of matter, the moral law, which is indestructible, belongs peculiarly to it—that is, to the spirit, and is inseparable from it.

We must next remember, that this earthly body we inhabit is more or less a mask, by means of which we conceal from each other those thoughts which, if constantly exposed, would unfit us for living in community; but when we die, this mask falls away, and the truth shows nakedly. There is no more disguise; we appear as we are, spirits of light or spirits of darkness; and there can be no difficulty, I should think, in conceiving this, since we know that even our present opaque and comparatively inflexible features, in spite of all efforts to the contrary, will be the index of the mind; and that the expression of the face is gradually moulded to the fashion of the thoughts. How much more must this be the case with the fluent and diaphanous body which we expect is to succeed the fleshly one!

Thus, I think, we have arrived at forming some conception of the state that awaits us hereafter; the indestructible moral law fixes our place or condition; affinity governs our associations; and the mask under which we conceal ourselves having fallen away, we appear to each other as we are: and I must here observe, that in this last circumstance, must be comprised one very important element of happiness or misery; for the love of the pure spirits for each other will be for ever excited by simply beholding that beauty and brightness which will be the inalienable expression of their goodness; whilst the reverse will be the case with the spirits of darkness; for no one loves wickedness, either in themselves or others, however we may practice it. We must also understand, that the words dark and light, which in this world of appearance we use metaphorically to express good and evil, must be understood literally when speaking of that other world where everything will be seen as it is. Goodness is truth, and truth is light; and wickedness is falsehood, and falsehood is darkness, and so it will be seen to be. Those who have not the light of truth to guide them will wander darkly through this valley of the shadow of death; those in whom the light of goodness shines will dwell in the light, which is inherent in themselves. The former will be in the kingdom of darkness, the latter in the kingdom of light. All the records existing of the blessed spirits that have appeared, ancient or modern, exhibit them as robed in light, whilst their anger or sorrow is symbolised by their darkness. Now, there appears to me nothing incomprehensible in this view of the future; on the contrary, it is the only one which I ever found myself capable of conceiving or reconciling with the justice and mercy of our Creator He does not punish us, we punish ourselves; we have built up a heaven or a hell to our own liking, and we carry it with us. The fire that for ever burns without consuming, is the fiery evil in which we have chosen our part; and the heaven in which we shall dwell will be the heavenly peace which will dwell in us. We are our own judges and our own chastisers; and here I must say a few words on the subject of that, apparently to us, preternatural memory which is developed under certain circumstances, and to which I alluded in a former chapter. Every one will have heard that persons who have been drowned and recovered have had, in what would have been their last moments, had no means been used to revive them, a strange vision of the past, in which their whole life seemed to float before them in review; and I have heard of the same phenomenon taking place in moments of impending death, in other forms. Now, as it is not during the struggle for life, but immediately before insensibility ensues, that this vision occurs, it must be the act of a moment; and this renders comprehensible to us what is said by the Seeress of Prevorst, and other somnambules of the highest order, namely, that the instant the soul is freed from the body it sees its whole earthly career in a single sign; it knows that it is good or evil, and pronounces its own sentence. The extraordinary memory occasionally exhibited in sickness where the link between the soul and the body is probably loosened, shows us an adumbration of this faculty.

But this self-pronounced sentence, we are led to hope is not final, nor does it seem consistent with the love and mercy of God that it should be so. There must be few, indeed, who leave this earth fit for heaven; for although the immediate frame of mind in which dissolution takes place, is probably very important, it is surely a pernicious error, encouraged by jail chaplains and philanthropists, that a late repentance and a few parting prayers can purify a soul sullied by years of wickedness. Would we at once receive such an one into our intimate communion and love?

Should we not require time for the stains of vice to be washed away and habits of virtue to be formed? Assuredly we should! And how can we imagine that the purity of heaven is to be sullied by that approximation that the purity of earth would forbid? It would be cruel to say, and irrational to think, that this late repentance is of no avail; it is doubtless so far of avail that the straining upwards and the heavenly aspirations of the parting soul are carried with it, so that when it is free, instead of choosing the darkness, it will flee to as much light as is in itself; and be ready, through the mercy of God and the ministering of brighter spirits, to receive more. But in this case, as also in the innumerable instances of those who die in what may be called a negative state, the advance must be progressive, though wherever the desire exists, I must believe that this advance is possible. If not, wherefore did Christ, after being "put to death in the flesh," go and "preach to the spirits in prison?" It would have been a mockery to preach salvation to those who had no hope; nor would they, having no hope, have listened to the preacher.

I think these views are at once cheering, encouraging, and beautiful; and I cannot but believe, that were they more generally entertained and more intimately conceived, they would be very beneficial in their effects. As I have said before, the extremely vague notions people have of a future life, prevent the possibility of its exercising any great influence upon the present. The picture, on one side, is too revolting and inconsistent with our ideas of Divine goodness to be deliberately accepted; whilst, with regard to the other, our feelings somewhat resemble those of a little girl, I once knew, who, being told by her mother what was to be the reward of goodness if she were so happy as to reach heaven, put her finger in her eye and began to cry, exclaiming, "Oh, mamma! how tired I shall be singing!"

The question which will now naturally arise, and which I am bound to answer, is, How have these views been formed? and what is the authority for them? and the answer I have to make will startle many minds, when I say, that they have been gathered from two sources; first and chiefly from the state in which those spirits appear to be, and sometimes avow themselves to be, who, after quitting the earth, return to it and make themselves visible to the living; and, secondly, from the revelations of numerous somnambules of the highest order, which entirely conform in all cases, not only with the revelations of the dead, but with each other. I do not mean to imply, when I say this, that I consider the question finally settled, as to whether somnambules are really clear-seers or only visionaries; nor that I have by any means established the fact that the dead do sometimes actually return; but I am obliged to beg the question for the moment, since whether these sources be pure or impure, it is from them the information has been collected. It is true, that these views are extremely conformable with those entertained by Plato and his school of philosophers; and also with those of the mystics of a later age; but the latter certainly, and the former probably, built up their systems on the same foundation; and I am very far from using the term *mystics* in the opprobrious, or at least contemptuous, tone in which it has of late years been uttered in this country; for although abounding in errors, as regarded the concrete, and although their want of an inductive methodology led them constantly astray in the region of the real, they were sublime teachers in that of the ideal; and they seem to have been endowed with a wonderful insight into this veiled department of our nature.

It may be here objected, that we only admire their insight, because, being in entire ignorance of the subject of it, we accept raving for revelation; and that no weight can be attached to the conformity of later disclosures with theirs, since they have no doubt been founded upon them. As to the ignorance, it is admitted; and, simply looking at their views, as they stand, they have nothing to support them but their sublimity and consistency; but, as regards the value of the evidence afforded by conformity, it rests on very different grounds; for the reporters from whom we collect our intelligence are, with very few exceptions, those of whom we may safely predicate, that they were wholly unacquainted with the systems promulgated by the Platonic philosophers, or the mystics either, nor, in most instances, had ever heard of their names; for, as regards that peculiar somnambulic state which is here referred to, the subjects of it appear to be generally very young people of either sex, and chiefly girls; and, as regards ghost-seeing, although this phenomenon seems to have no connexion with the age of the seer, yet it is not usually from the learned or the cultivated we collect our cases, inasmuch as the apprehension of ridicule, on the one hand, and the fast hold the doctrine of spectral illusions has taken of them, on the other, prevent their believing in their own senses, or producing any evidence they might have to furnish.

And here will be offered another subtle objection, namely, that the testimony of such witnesses as I have above described is perfectly worthless; but this I deny. The somnambulic states I allude to, are such as have been developed, not artificially, but naturally; and often under very extraordinary nervous diseases, accompanied with catalepsy, and various symptoms far beyond feigning. Such cases are rare, and, in this country, seem to have been very little observed, for doubtless they must occur, and when they do occur, they are very carefully concealed by the families of the patient, and not followed up or investigated as a psychological phenomenon by the physician; for it is to be observed that, without questioning no revelations are made; they are not, as far as I know, ever spontaneous. I have heard of two such cases in this country, both occurring in the higher classes, and both patients being young ladies; but, although surprising phenomena were exhibited, interrogation was not permitted, and the particulars were never allowed to transpire.

No doubt there are examples of error and examples of imposture, so there are in everything where room is to be found for them; and I am quite aware of the propensity of hysterical patients to deceive, but it is for the judicious observers to examine the genuineness of each particular instance; and it is perfectly certain and well established by the German physiologists and psychologists, who have carefully studied the subject, that there are many above all suspicion. Provided, then, that the case be genuine, it remains to be determined how much value is to be attached to the revelations, for they may be quite honestly delivered, and yet be utterly worthless—the mere ravings of a disordered brain; and it is here that conformity becomes important, for I cannot admit the objection that the simple circumstance of. the patient's being diseased invalidates their evidence so entirely as to annul even the value of their unanimity, because although it is not logically impossible, that a certain state of nervous derangement should occasion all somnambules, of the class in question, to make similar answers, when interrogated, regarding a subject of which in their normal condition they know nothing, and on which they have never reflected, and that these answers should be not only consistent, but disclosing far more elevated views than are evolved by minds of a very superior order which *have* reflected on it very deeply—I say, although this is not logically impossible, it will assuredly be found, by most persons, an hypothesis of much more difficult acceptance than the one I propose; namely, that whatever be the cause of the effect, these patients are in a state of clear-seeing, wherein they have "more than mortal knowledge;" that is, more knowledge than mortals possess in their normal condition: and it must not be forgotten, that we have some facts confessed by all experienced physicians and physiologists, even in this country, proving that there are states of disease in which preternatural faculties have been developed, such as no theory has yet satisfactorily accounted for.

But Dr. Passavent, who has written a very philosophical work on the subject of vital magnetism and clear-seeing, asserts, that it is an error to imagine that the extatic condition is merely the product of disease. He says, that it has sometimes exhibited itself in persons of very vigorous constitutions, instancing Joan of Arc, a woman, whom historians have little understood, and whose memory Voltaire's detestable poem has ridiculed and degraded, but who was, nevertheless, a great psychological phenomenon.

The circumstance, too, that phenomena of this kind are more frequently developed in women than in men, and that they are merely the consequence of her greater nervous irritability has been made another objection to them—an objection, however, which Dr. Passavent considers founded on ignorance of the essential difference between the sexes, which is not merely a physical but a psychological one. Man is more productive than receptive. In a state of perfectibility, both attributes would be equally developed in him; but in this terrestrial life, only imperfect phases of the entire sum of the soul's faculties are so. Mankind are but children, male or female, young or old: of man, in his totality, we have but faint adumbrations, here and there.

Thus the extatic woman will be more frequently a seer, instinctive and intuitive; man, a doer and a worker; and as all genius is a degree of extacy or clear-seeing, we perceive the reason wherefore in man it is more productive than in woman, and that our greatest poets and artists, in all kinds, are of the former sex, and even the most remarkable women produce but little in science or art; whilst on the other hand, the feminine instinct, and tact, and intuitive seeing of truth, is frequently more sure than the ripe and deliberate judgment of man: and it is hence that solitude and such conditions as develop the passive or receptive at the expense of the active, tend to produce this state, and to assimilate the man more to the nature of the woman; whilst in her they intensify these distinguishing characteristics: and this is also the reason that simple and child-like people and races are the most frequent subjects of these phenomena.

It is only necessary to read Mozart's account of his own moments of inspiration, to comprehend, not only the similarity, but the positive identity of the extatic state with the state of genius in activity. "When all goes well with me," he says, "when I am in a carriage, or walking, or when I cannot sleep at night, the thoughts come streaming in upon me most fluently. Whence, or how, is more than I can tell. What comes, I hum to myself, as it proceeds................then follows the counterpoint and the clang of the different instruments, and if I am not disturbed my soul is fixed, and the thing grows greater, and broader, and clearer; and I have it all in my head, even when the piece is a long one, and I see it like a beautiful picture, not hearing the different parts in succession, as they must be played, but the whole at once. That is the delight! The composing and the making is like a beautiful and vivid dream, but this hearing of it, is the best of all."

What is this but clear-seeing, backwards and forwards, the past and the future? The one faculty is not a whit more surprising and incomprehensible than the other, to those who possess neither, only we see the material product of one, and therefore believe in it. But, as Passavent justly says, these corruscations belong not to genius exclusively: they are latent in all men. In the highly gifted, this divine spark becomes a flame to light the world withal: but even in the coarsest and least developed organizations, it may, and does momentarily break forth. The germ of the highest spiritual life is in the rudest, according to its degree, as well as in the highest form of man we have yet seen; he is but a more imperfect type of the race, in whom this spiritual germ has not unfolded itself.

Then, with respect to our second source of information, I am quite aware that it is equally difficult to establish its validity; but there are a few arguments in our favour here, too. In the first place, as Dr. Johnson says, though all reason is against us, all tradition is for us; and this conformity of tradition is surely of some weight, since I think it would be difficult to find any parallel instance, of a universal tradition that was entirely without a foundation in truth; for with respect to witchcraft, the belief in which is equally universal, we now know that the phenomena were generally facts, although the interpretations put upon them were fables. It may certainly be objected that this universal belief in ghosts only arises from the universal prevalence of spectral illusions, but, if so, as I have before observed, these spectral illusions become a subject of very curious enquiry, for, in the first place, they frequently occur under circumstances the least likely to induce them, and to people whom we should least expect to find the victims of them; and, in the second, there is a most remarkable conformity here, too, not only between the individual cases occurring amongst all classes of persons, who had never exhibited the slightest tendency to nervous derangement or somnambulism, but also between these and the revelations of the somnambules. In short, it seems to me that life is reduced to a mere phantasmagoria, if spectral illusions are so prevalent, so complicated in their nature, and so delusive as they must be, if all the instances of ghost-seeing that come before us are to be referred to that theory. How numerous these are, I confess myself not to have had the least idea, till my attention was directed to the enquiry; and that these instances have been equally frequent in all periods and places, we cannot doubt, from the variety of persons that have given in their adhesion, or at least that have admitted, as Addison did, that he could not refuse the universal testimony in favour of the re-appearance of the dead, strengthened by that of many credible persons with whom he was acquainted. Indeed, the testimony in favour of the facts has been at all periods too strong to be wholly rejected, so that even the materialists, like Lucretius and the elder Pliny, find themselves obliged to acknowledge them, whilst, on the other hand, the extravagant admissions that are demanded of us by those who endeavour to explain them away, prove that their disbelief rests on no more solid foundation than their own prejudices. I acknowledge all the difficulty of establishing the facts, such difficulties as indeed encompass few other branches of enquiry; but I maintain that the position of the opponents is still worse, although, by their high tone, and their contemptuous laugh, they assume to have taken up one that, being fortified by reason, is quite impregnable, forgetting that the wisdom of man is preeminently "foolishness before God," when it wanders into this region of unknown things. Forgetting, also, that they are just serving this branch of enquiry, as their predecessors, whom they laugh at, did physiology; concocting their systems out of their own brains, instead of the responses of nature; and with still more rashness and presumption, this department of her kingdom being more inaccessible, more incapable of demonstration, and more entirely beyond our controul; for these spirits will not "come when we do call them;" and, I confess, it often surprises me to hear the very shallow nonsense that very clever men talk upon the subject, and the inefficient arguments they use to disprove what they know nothing about. I am quite conscious that the facts I shall adduce are open to controversy; I can bring forward no evidence that will satisfy a scientific mind; but neither are my opponents a whit better fortified. All I do hope to establish is, not a proof, but a presumption; and the conviction I desire to awaken in people's minds, is, not that these things *are* so, but that they *may* be so, and that it is well worth our while to enquire whether they are or not.

It will be seen, that these views of a future state are extremely similar to those of Isaac Taylor, as suggested in his physical theory of another life—at least, as far as he has entered upon the subject—and it is natural that they should be so, because he seems also to have been a convert to the opinion, that "the dead do sometimes break through the boundaries that hem in the etherial crowds; and if so, as if by trespass, may in single instances infringe upon the ground of common corporeal life."

Let us now fancy this dispossessed soul entering on its new career, amazed, and no more able than when it was in the body to accommodate itself at once to conditions of existence, for which it was unprepared. If its aspirations had previously been heavenward, these conditions would not be altogether new, and it would speedily find itself at home in a sphere in which it had dwelt before; for, as I have formerly said, a spirit must be where its thoughts and affections are. and the soul, whose thoughts and affections had been directed to heaven, would only awaken after death into a more perfect and unclouded heaven. But imagine the contrary of all this. Conceive what this awakening must be to an earth-bound spirit—to one altogether unprepared for its new home—carrying no light within it—floating in the dim obscure—clinging to the earth, where all its affections were garnered up; for where its treasure is, there shall it be also. It will find its condition evil, more or less, according to the degree of its moral light or darkness, and in proportion to the amount of the darkness will be its incapacity to seek for light. Now, there seems nothing offensive to our notions of the Divine goodness in this conception of what awaits us when the body dies. It appears to me, on the contrary, to offer a moreoc mprehensible and coherent view than any other that has been presented to me; yet, the state I have depicted is very much the Ha des of the Greeks and Romans. It is the middle state, on which all souls enter, a state in which there are many mansions—that is, there are innumerable states—probably not permanent, but ever progressive or retrograde; for we cannot conceive of any moral state being permanent, since we know perfectly well that ours is never so: it is always advancing or retroceding. When we are not improving, we are deteriorating; and so it must necessarily be with us hereafter.

Now, if we admit the probability of this middle state, we have removed one of the great objections which are made to the belief in the re-appearance of the dead; namely, that the blest are too happy to return to the earth, and that the wicked have it not in their power to do so. This difficulty arises, however, very much from the material ideas entertained of Heaven and Hell—the notion that they are places instead of states. I am told that the Greek word *Hades* is derived from *æides, invisible;* and that the Hebrew word *Scheôl*, which has the same signification, also implies a state, not a place; since it may be interpreted into *desiring, longing, asking, praying.* These words in the Septuagint, are translated by *grave, death,* or *hell;* but previously to the Reformation, they seem to have borne their original meaning; that is, the state into which the soul entered at the death of the body. It was probably to get rid of the purgatory of the Roman Church, which had doubtless become the source of many absurd notions and corrupt practices, that the doctrine of a middle state or Hades was set aside; besides which the honest desire for reformation in all reforming churches, being alloyed by the *odium theologicum*, the purifying besom is apt to take too discursive a sweep, exercising less modesty and discrimination than might be desirable; and thus not uncommonly wiping away truth and falsehood together.

Dismissing the idea, therefore, that Heaven and Hell are places in which the soul is imprisoned, whether in bliss or woe, and, supposing that, by a magnetic relation, it may remain connected with the sphere to which it previously belonged, we may easily conceive that, if it have the memory of the past, the more entirely sensuous its life in the body may have been, the closer it will cling to the scene of its former joys; or, even if its sojourn on earth were not a period of joy, but the contrary, still, if it have no heavenward aspirations, it will find itself, if not in actual woe, yet aimless, objectless, and out of a congenial element. It has no longer the organs whereby it perceived, communicated with, and enjoyed the material world and its pleasures. The joys of Heaven are not its joys; we might as well expect a hardened prisoner in Newgate, associating with others as hardened as himself, to melt into extatic delight at the idea of that which he cannot apprehend! How helpless and inefficient such a condition seems, and how natural it is to us to imagine that, under such circumstances, there might be awakened a considerable desire to manifest itself to those yet living in the flesh, if such a manifestation be possible! And what right have we, in direct contradiction to all tradition, to assert that it is not? We may raise up a variety of objections from physical science, but we cannot be sure that these are applicable to the case; and of the laws of spirit we know very little, since we are only acquainted with it as circumscribed, confined, and impeded in its operations by the body; and whenever such abnormal states occur as enable it to act with any degree of independence, man, under the dominion of his all-sufficient reason, denies and disowns the facts. That the manifestation of a spirit to the living, whether seen or heard, is an exception, and not the rule, is evident; for, supposing the desire to exist at all, it must exist in millions and millions of instances which never take effect. The circumstances must, therefore, no doubt be very peculiar, as regards both parties in which such a manifestation is possible; what these are we have very little means of knowing, but, as far as we do know, we are led to conclude that a certain magnetic rapport or polarity constitute this condition, whilst, at the same time, as regards the seer, there must be what the prophet called the "*opening of the eye*," which may, perhaps, signify the seeing of the spirit without the aid of the bodily organ, a condition which may temporarily occur to any one under we know not what influence, but which seems, to a certain degree, hereditary in some families.

The following passage is quoted from Sir William Hamilton's edition of Dr. Reid's works, published in 1846:—

"No man can show it to be impossible to the Supreme Being to have given us the power of perceiving external objects, without any such organs" *i. e.*, our organs of sense. "We have reason to believe that when we put off these bodies, and all the organs belonging to them, our perceptive powers shall rather be improved than destroyed or impaired. We have reason to believe that the Supreme Being perceives everything in a much more perfect manner than we do, without bodily organs. We have reason to believe that there are other created beings endowed with powers of perception more perfect and more extensive than ours, without any such organs as we find necessary;" and Sir William Hamilton adds the following note:—

"However astonishing, it is now proved beyond all rational doubt, that in certain abnormal states of the nervous organism, perceptions are possible through other than the ordinary channels of the senses."

Of the existence of this faculty in nature, any one, who chooses, may satisfy himself by a very moderate degree of trouble, provided he undertake the investigation honestly; and this being granted, another objection, if not altogether removed, is considerably weakened. I allude to the fact, that in numerous reported cases of ghost-seeing, the forms were visible to only one person, even though others were present, which, of course, rendered them undistinguishable from cases of spectral illusion, and indeed unless some additional evidence be afforded, they must remain so still, only we have gained thus much, that this objection is no longer unanswerable; for whether the phenomenon is to be referred to a mutual rapport, or to the opening of the spiritual eye, we comprehend how one may see what others do not. But really, if the seeing depended upon ordinary vision, I cannot perceive that the difficulty is insurmountable; for we perfectly well know that some people are endowed with an acuteness of sense, or power of perception, which is utterly incomprehensible to others: for without entering into the disputed region of clear-seeing, everybody must have met with instances of those strange antipathies to certain objects, accompanied by an extraordinary capacity for perceiving their presence, which remain utterly unexplained. Not to speak of cats and hares, where some electrical effects might be conceived, I lately heard of a gentleman who fainted if he were introduced into a room where there was a raspberry tart; and that there have been persons endowed with a faculty for discovering the proximity of water and metals, even without the aid of the divining rod—which latter marvel seems to be now clearly established as an electrical phenomenon, will scarcely admit of further doubt. A very eminent person, with whom I am acquainted, possessing extremely acute olfactory powers, is the subject of one single exception. He is insensible to the odour of a bean-field, however potent: but it would surely be very absurd in him to deny that the bean-field emits an odour, and the evidence of the majority against him is too strong to admit of his doing so. Now, we have only the evidence of a minority with regard to the existence of certain faculties not generally developed, but surely it argues great presumption to dispute their possibility. We might, I think, with more appearance of reason, insist upon it that my friend *must* be mistaken, and that he does smell the bean-field; for we have the majority against him there, most decidedly. The difference is, that nobody cares whether the odour of the bean-field is perceptible or not: but if the same gentleman asserted that he had seen a ghost, beyond all doubt, his word would be disputed.

Though we do not know what the conditions are that develope the faculty of what St. Paul calls the discerning of spirits; there is reason to believe that the approach of death is one. I have heard of too many instances of this kind, where the departing person has been in the entire possession of his or her faculties, to doubt that in our last moments we are frequently visited by those who have gone before us, and it being admitted by all physiologists, that preternatural faculties are sometimes exhibited at this period, we can have no right to say that "the discerning of spirits" is not one of them.

There is an interesting story recorded by Beaumont, in his "World of Spirits," and quoted by Dr. Hibbert with the remark, that no reasonable doubt can be placed on the authenticity of the narrative, as it was drawn up by the Bishop of Gloucester from the recital of the young lady's father; and I mention it here not for any singularity attending it, but first because its authenticity is admitted, and next on account of the manner in which, so much being granted, the fact is attempted to be explained away.

"Sir Charles Lee, by his first lady, had only one daughter, of which she died in child-birth, and when she was dead, her sister, the Lady Everard, desired to have the education of the child, and she was very well educated till she was marriageable, and a match was concluded for her with Sir W. Parkins, but was then prevented in an extraordinary manner. Upon a Thursday night, she thinking she saw a light in her chamber after she was in bed, knocked for her maid, who presently came to her, and she asked, 'Why she left a candle burning in her room?' The maid answered, that she had 'left none, and that there was none but what she had brought with her at that time;' then, she said, it must be the fire; but that her maid told her, was quite out, adding she believed it was only a dream, whereupon Miss Lee answered, it might be so, and composed herself again to sleep. But, about two of the clock, she was awakened again, and saw the apparition of a little woman between her curtains and her pillow, who told her she was her mother, that she was happy, and that, by twelve of the clock that day, she should be with her. Whereupon, she knocked again for her maid, called for her clothes, and when she was dressed, went into her closet, and came not out again till nine, and then brought out with her a letter, sealed, to her father, carried it to her aunt, the Lady Everard, told her what had happened, and desired that as soon as she was dead it might be sent to him. The lady thought she was suddenly fallen mad, and therefore sent presently away to Chelmsford, for a physician and surgeon, who both came immediately, but the physician could discern no indication of what the lady imagined, or of any indisposition of her body; notwithstanding, the lady would needs have her let blood, which was done accordingly; and when the young woman had patiently let them do what they would with her, she desired that the chaplain might be called to read prayers; and when prayers were ended, she took her guitar and psalm-book, and sat down upon a chair without arms, and played and sung so melodiously and admirably, that her music-master, who was then there, admired at it; and near the stroke of twelve, she rose and sat herself down in a great chair with arms, and presently fetching a strong breathing or two, she immediately expired, and was so suddenly cold as was much wondered at by the physician and surgeon. She died at Waltham, in Essex, three miles from Chelmsford, and the letter was sent to Sir Charles, at his house, in Warwickshire; but he was so afflicted at the death of his daughter, that he came not till she was buried: but when he came, he caused her to be taken up, and to be buried with her mother, at Edmonton, as she desired in her letter."

This circumstance occurred in the year 1662, and is, as Dr. Hibbert observes, "one of the most interesting ghost-stories on record:" yet he insists on placing it under the category of spectral illusions, upon the plea, that let the physician, whose skill he arraigns, say what he would, her death within so short a period,.

proves that she must have been indisposed at the time she saw the vision, and that probably "the languishing female herself might have unintentionally contributed to the more strict verification of the ghost's prediction," concluding with these words, "all that can be said of it is, that the coincidence was a *fortunate one;* for without it, the story would, probably, never have met with a recorder,' &c. &c.

Now, I ask if this is a fair way of treating any fact, transmitted to us on authority, which the objector himself admits to be perfectly satisfactory; more especially, as the assistants on the occasion appear to have been quite as unwilling to believe in the *supernatural* interpretation of it, as Dr. H. could have been himself, had he been present; for what more could he have done than conclude the young lady to be mad, and bled her?—a line of practice which is precisely what would be followed at the present time; and which proves that they were very well aware of the sensuous illusions produced by a disordered state of the nervous system; and with respect to his conclusion that the "languishing female" contributed to the verification of the prediction, we are entitled to ask, where is the proof that she was languishing? A very clever watchmaker once told me, that a watch may go perfectly well for years and at length stop suddenly, in consequence of an organic defect in its construction, which only becomes perceptible, even to the eye of a watchmaker, when this effect takes place; and we do know that many persons have suddenly fallen dead immediately after declaring themselves in the best possible health; and we have therefore no right to dispute what the narrator implies, namely, that there were no sensible indications of the impending catastrophe.

There either was some organic defect or derangement in this lady's physical economy, which rendered her death inevitable at the hour of noon, on that particular Thursday, or there was not. If there were, and her certain death was impending at that hour, how came she acquainted with the fact? Surely, it is a monstrous assumption to say, that it was "a fortunate coincidence," when no reason whatever is given us for concluding that she felt otherwise than perfectly well? If, on the contrary, we are to take refuge in the supposition that there was no death impending, and that she only died of the fright, how came she—feeling perfectly well, and, in this case, we have a right to conclude *being* perfectly well,—to be the subject of such an extraordinary spectral illusion? And if such spectral illusions can occur to people in a good normal state of health, does it not become very desirable to give us some clearer theory of them than we have at present. But there is a third presumption to which the sceptical may have recourse, in order to get rid of this well established, and therefore very troublesome fact, namely, that Miss Lee *was* ill, although unconscious of it herself, and indicating no symptoms that could guide her physician to an enlightened diagnosis; and that the proof of this is to be found in the occurrence of the spectral illusion, and that this spectral illusion so impressed her, that it occasioned the precise fulfilment of the imaginary prediction, an hypothesis which appears to me to be pressing very hard on the spectral illusion; for it is first called upon to establish the fact of an existing indisposition of no slight character, of which neither patient or physician were aware; and it is next required to kill the lady with unerring certainty, at the hour appointed, she being, according to the only authority we have for the story, in a perfectly calm and composed state of mind! for there is nothing to be discerned in the description of her demeanour but an entire and willing submission to the announced decree, accompanied by that pleasing exaltation, which appears to me perfectly natural under the circumstances; and I do not think that anything we know of human vitality can justify us in believing that life can be so easily extinguished. But to such straights people are reduced, who write with a predetermination to place their facts on a Procrustian bed, till they have fitted them into their own cherished theory.

In the above recorded case of Miss Lee, the motive for the visit is a sufficient one; but one of the commonest objections to such narrations, is the insignificance of the motive when any communication is made, or there being apparently no motive at all, when none is made. Where any previous attachment has subsisted, we need seek no further for an impelling cause; but, in other cases, this impelling cause must probably be sought in the earthly rapport still subsisting and the urgent desire of the spirit to manifest itself and establish a communication where its thoughts and affections still reside; and we must consider that, provided there be no law of God prohibiting its revisiting the earth, which law would of course supersede all other laws, then, as I have before observed, where its thoughts and affections are, it must be also. What is it but our heavy material bodies that prevents us from being where our thoughts are? But the being near us, and the manifesting itself to us, are two very different things, the latter evidently depending on conditions we do not yet understand. As I am not writing a book on vital magnetism, and there are so many already accessible to every body who chooses to be informed on it, I shall not here enter into the subject of *magnetic rapport*, it being, I believe, now generally admitted, except by the most obstinate sceptics, that such a relation can be established betwixt two human beings. In what this relation consists, is a more difficult question, but the most rational view appears to be that of a magnetic polarity which is attempted to be explained by two theories—the dynamical and the etherial: the one viewing the phenomena as simply the result of the transmission of forces, the other hypothetising an ether which pervades all space, and penetrates all substance, maintaining the connexion betwixt body and soul, and betwixt matter and spirit. To most minds, this last hypothesis will be the most comprehensible; on which account, since the result would be the same in either case, we may adopt it for the moment; and there will then be less difficulty in conceiving that the influence or ether of every being or thing, animate or inanimate, must extend beyond the periphery of its own terminations: and that this must be eminently the case where there is animal life, the nerves forming the readiest conductors for this supposed imponderable. The proofs of the existence of this ether are said to be manifold, and more especially to be found in the circumstances that every created thing sheds an atmosphere around it, after its kind; this atmosphere becoming, under certain conditions, perceptible or even visible, as in the instances of electric fish, &c., the fascinations of serpents, the influence of human beings upon plants, and *vice versa;* and finally, the phenomena of animal magnetism, and the undoubted fact, to which I myself can bear witness, that the most ignorant girls, when in a state of somnambulism, have been known to declare that they saw their magnetiser surrounded by a halo of light; and it is doubtless this halo of light, that, from their being strongly magnetic men, has frequently been observed to surround the heads of saints and eminently holy persons: the temperament that produced the internal fervour, causing the visible manifestation of it. By means of this ether, or force, a never-ceasing motion and an intercommunication is sustained betwixt all created things, and betwixt created things and their Creator, who sustains them and creates them ever anew, by the constant exertion of his Divine will, of which this is the messenger and the agent, as it is betwixt our will and our own bodies; and without this sustaining will, so exerted, the whole would fall away, dissolve and die; for it is the life of the universe. That all inanimate objects emit an influence, greater or less, extending beyond their own peripheries is established by their effects on various susceptible individuals, as well as on somnambules; and thus there exists a universal polarity and rapport, which is however stronger betwixt certain organisms; and every being stands in a varying relation of positive and negative to every other.

With regard to these theories, however, where there is so much obscurity, even in the language, I do not wish to insist; more especially as I am fully aware that this subject may be discussed in a manner much more congruous with the dynamical spirit of the philosophy of this century: but, in the meanwhile, as either of the causes alluded to is capable of producing the effects, we adopt the hypothesis of an all-pervading ether, as the one most easily conceived.

Admitting this then to be the case, we begin to have some notion of the *modus operandi*, by which a spirit may manifest itself to us, whether to our internal universal sense, or even to our sensuous organs; and we also find one stumbling block removed out of our way, namely, that it shall be visible or even audible to one person and not to another, or at one time and not at another; for by means of this ether, or force, we are in communication with all spirit, as well as with all matter; and since it is the vehicle of will, a strong exertion of will may reinforce its influence to a degree far beyond our ordinary conceptions: but man is not acquainted with his own power, and has consequently no faith in his own will: nor is it probably the design of Providence, in ordinary cases, that he should. He cannot therefore exert it; if he could, he "might remove mountains." Even as it is, we know something of the power of will in its effect on other organisms, as exhibited by certain strong-willed individuals; also in popular movements, and more manifestly in the influence and far-working of the magnetiser on his patient. The power of will, like the seeing of the spirit, is latent in our nature, to be developed in God's own time; but meanwhile, slight examples are found, shooting up here and there, to keep alive in man the consciousness that he is a spirit, and give evidence of his divine origin.

What especial laws may appertain to this supersensuous domain of nature, of course we cannot know, and it is therefore impossible for us to pronounce how far a spirit is free, or not free, at all times to manifest itself; and we can, therefore, at present, advance no reason for these manifestations not being the rule instead of the exception. The law which restrains more frequent intercourse, may, for anything we know to the contrary, have its relaxations and its limitations, founded in nature; and a rapport with, or the power of acting on, particular individuals, may arise from causes of which we are equally ignorant. Undoubtedly, the receptivity of the corporeal being is one of the necessary conditions, whilst, on the part of the incorporeal, the will is at once the cause and the agent that produces the effect; whilst attachment, whether to individuals or to the lost joys of this world, is the motive. The happy spirits in whom this latter impulse is weak, and who would float away into the glorious light of the pure moral law, would have little temptation to return; and at least would only be brought back by their holy affections or desire to serve mankind. The less happy, clinging to their dear corporeal life, would hover nearer to the earth; and I do question much whether the often ridiculed idea of the mystics, that there is a moral *weight*, as well as a moral *darkness*, be not founded in truth. We know very well that even these substantial bodies of ours, are, to our own sensations (and, very possibly, if the thing could be tested, would prove to be in fact) lighter or heavier, according to the lightness or heaviness of the spirit—terms used figuratively, but perhaps capable of a literal interpretation; and thus the common idea of *up* and *down*, as applied to Heaven or Hell, is founded in truth, though not mathematically correct, we familiarly using the words up and down to express *farther* or *nearer*, as regards the planet on which we live.

Experience seems to justify this view of the case; for, supposing the phenomena I am treating of to be facts, and not spectral illusions, all tradition shows that the spirits most frequently manifested to man, have been evidently not in a state of bliss; whilst, when bright ones appeared, it has been to serve him; and hence the old persuasion that they were chiefly the wicked that haunted the earth, and hence, also, the foundation for the belief that not only the murderer, but the murdered, returned to vex the living; and the just view, that in taking away life the injury is not confined to the body, but extends to the surprised and angry soul, which is—

> "Cut off, even in the blossom of its sin,
> Unhousel'd, disappointed, unaneal'd;
> No reckoning made, but sent to its account
> With all its imperfections on its head."

It seems also to be gathered from experience, that those whose lives have been rendered wretched, "rest not in their graves," at least, several accounts I have met with, as well as tradition, countenance this view; and this may originate in the fact, that cruelty and ill-usage frequently produce very pernicious effects on the mind of the sufferer, in many instances inspiring, not resignation or a pious desire for death, but resentment, and an eager longing for a fair share of earthly enjoyment.

Supposing, also, the feelings and prejudices of the earthly life to accompany this dispossessed soul—for, though the liberation from the body inducts it into certain privileges inherent in spirit, its moral qualities remain as they were, as the tree falls, so it shall lie—supposing, therefore, that these feelings, and prejudices, and recollections of its past life, are carried with it, we see, at once, why the discontented spirits of the Heathen world could not rest till their bodies had obtained sepulture, why the buried money should torment the soul of the miser, and why the religious opinions, whatever they may have been, believed in the flesh, seem to survive with the spirit. There are two remarkable exceptions, however, and these are precisely such as might be expected. Those who, during their corporeal life, have not believed in a future state, return to warn their friends against the same error. "There is another world," said the brother of the young lady who appeared to her in the Cathedral of York, on the day he was drowned; and there are several similar instances recorded. The belief that this life "is the be-all and the end-all here," is a mistake that death must instantly rectify. The other exception I allude to is, that that toleration, of which, unfortunately, we see much less than is desirable in this world, seems happily to prevail in the next; for, amongst the numerous narrations I meet with, in which the dead have returned to ask the prayers or the services of the living, they do not seem, as will be seen by and by, to apply by any means exclusively, to members of their own church. The *attrait* which seems to guide their selection of individuals, is evidently not of a polemical nature. The pure worship of God, and the inexorable moral law, are what seem to prevail in the other world, and not the dogmatic theology which makes so much of the misery of this.

There is a fundamental truth in all religions; the real end of all is morality, however the means may be mistaken, and however corrupt, selfish, ambitious, and sectarian the mass of their teachers may, and generally do, become; whilst the effect of prayer, in whatever form, or to whatever ideal of the Deity, it may be offered, provided that offering be honestly and earnestly made, is precisely the same to the supplicant and in its results.

I have reserved the following story, which is not a fiction, but the relation of an undoubted and well-attested fact, till the present chapter, as being particularly applicable to this branch of my subject.

Some ninety years ago, there flourished in Glasgow a club of young men, which, from the extreme profligacy of its members and the licentiousness of their orgies, was commonly called the Hell Club. Besides their nightly or weekly meetings, they held one grand annual saturnalia, in which each tried to excel the other in drunkenness and blasphemy ; and on these occasions there was no star amongst them whose lurid light was more conspicuous than that of young Mr. Archibald B., who, endowed with brilliant talents and a handsome person, had held out great promise in his boyhood, and raised hopes, which had been completely frustrated by his subsequent reckless dissipations.

One morning, after returning from this annual festival, Mr. Archibald B. having retired to bed, dreamt the following dream:—

He fancied that he himself was mounted on a favourite black horse, that he always rode, and that he was proceeding towards his own house, then a country seat embowered by trees, and situated upon a hill, now entirely built over, and forming part of the city, when a stranger, whom the darkness of night prevented his distinctly discerning, suddenly seized his horse's rein, saying, "You must go with me!"

"And who are you?" exclaimed the young man, with a volley of oaths, whilst he struggled to free himself.

"That you will see by and by," returned the other, in a tone that excited unaccountable terror in the youth, who plunging his spurs into his horse, attempted to fly. But in vain: however fast the animal flew, the stranger was still beside him, till at length in his desperate efforts to escape, the rider was thrown, but instead of being dashed to the earth, as he expected, he found himself falling—falling—falling still, as if sinking into the bowels of the earth.

At length, a period being put to this mysterious descent, he found breath to enquire of his companion, who was still beside him, whither they were going; "Where am I? Where are you taking me?" he exclaimed.

"To Hell!" replied the stranger, and immediately interminable echoes repeated the fearful sound, "To Hell! to Hell! to Hell!"

At length a light appeared, which soon increased to a blaze; but, instead of the cries, and groans, and lamentings, the terrified traveller expected, nothing met his ear but sounds of music, mirth, and jollity; and he found himself at the entrance of a superb building, far exceeding any he had seen constructed by human hands. Within, too, what a scene! No amusement, employment, or pursuit of man on earth, but was here being carried on with a vehemence that excited his unutterable amazement. "There the young and lovely still swam through the mazes of the giddy dance! There the panting steed still bore his brutal rider through the excitements of the goaded race! There, over the midnight bowl, the intemperate still drawled out the wanton song or maudlin blasphemy! The gambler plied for ever his endless game, and the slaves of Mammon toiled through eternity their bitter task; whilst all the magnificence of earth paled before that which now met his view!"

He soon perceived that he was amongst old acquaintance, whom he knew to be dead, and each he observed, was pursuing the object, whatever it was, that had formerly engrossed him; when, finding himself relieved of the presence of his unwelcome conductor, he ventured to address his former friend Mrs. D., whom he saw sitting, as had been her wont on earth, absorbed at loo, requesting her to rest from the game, and introduce him to the pleasures of the place, which appeared to him to be very unlike what he had expected, and, indeed, an extremely agreeable one. But, with a cry of agony, she answered, that there was no rest in Hell; that they must ever toil on at those very pleasures; and innumerable voices echoed through the interminable vaults, "There is no rest in Hell!" Whilst, throwing open their vests, each disclosed in his bosom an ever-burning flame! These, they said, were the pleasures of Hell; their choice on earth was now their inevitable doom! In the midst of the horror this scene inspired, his conductor returned, and, at his earnest entreaty, restored him again to earth; but, as he quitted him, he said, "Remember! In a year and a day we meet again!"

At this crisis of his dream, the sleeper awoke, feverish and ill; and, whether from the effect of the dream, or of his preceding orgies, he was so unwell as to be obliged to keep his bed for several days, during which period he had time for many serious reflections, which terminated in a resolution to abandon the club and his licentious companions altogether.

He was no sooner well, however, than they flocked around him, bent on recovering so valuable a member of their society; and having wrung from him a confession of the cause of his defection, which, as may be supposed, appeared to them eminently ridiculous, they soon contrived to make him ashamed of his good resolutions. He joined them again, resumed his former course of life, and when the annual saturnalia came round, he found himself with his glass in his hand at the table, when the president, rising to make the accustomed speech, began with saying, "Gentleman: This being leap-year, it is a year and a day since our last anniversary, &c. &c." The words struck upon the young man's ear like a knell; but ashamed to expose his weakness to the jeers of his companions, he sat out the feast, plying himself with wine, even more liberally than usual, in order to drown his intrusive thoughts; till, in the gloom of a winter's morning, he mounted his horse to ride home. Some hours afterwards, the horse was found with his saddle and bridle on, quietly grazing by the road-side, about half-way between the city and Mr. B.'s house; whilst a few yards off, lay the corpse of his master.

Now, as I have said, in introducing this story, it is no fiction: the circumstance happened as here related. An account of it was published at the time, but the copies were bought up by the family. Two or three however were preserved, and the narrative has been re-printed.

The dream is evidently of a symbolical character; and accords in a very remarkable degree with the conclusions to be drawn from the sources I have above indicated. The interpretation seems to be, that the evil passions and criminal pursuits which have been indulged in here become our curse hereafter. I do not mean to imply that the ordinary amusements of life are criminal; far from it. There is no harm in dancing, nor in playing at loo, either; but if people make these things the whole business of their lives, and think of nothing else, cultivating no higher tastes, nor forming no higher aspirations, what sort of preparation are they making for another world? I can hardly imagine that anybody would wish to be doing these things to all eternity, the more especially that it is most frequently *ennui* that drives their votaries into excesses, even here; but if they have allowed their minds to be entirely absorbed in such frivolities and trivialties, surely they cannot expect that God will, by a miracle, suddenly obliterate these tastes and inclinations, and inspire them with others better suited to their new condition! It was their business to do that for themselves, whilst here; and such a process of preparation is not in the slightest degree inconsistent with the enjoyment of all manner of harmless pleasures; on the contrary, it gives the greatest zest to them; for a life, in which there is nothing serious, in which all is play and diversion, is, beyond all doubt, next to a life of active, persevering wickedness, the saddest thing under the sun! But let everybody remember, that we see in nature no violent transitions; everything advances by almost insensible steps, at least everything that is to endure, and therefore to expect that because they have quitted their fleshly bodies, which they always knew were but a temporary appurtenance, doomed to perish and decay, they themselves are to undergo a sudden and miraculous conversion and purification, which is to elevate them into fit companions for the angels of Heaven, and the Blessed that have passed away, is surely one of the most inconsistent, unreasonable, and pernicious errors that mankind ever indulged in!

APPENDIX TO CHAPTER VI.

CASE OF COLONEL TOWNSHEND.

Whilst this volume is going through the press, I find, from the account of Dr. Cheyne, who attended him, that Colonel Townshend's own way of describing the phenomenon to which he was subject, was, that he could "die or expire when he pleased; and yet, by an effort, or *somehow,* he could come to life again." He performed the experiment in the presence of three medical men, one of whom kept his hand on his heart, another held his wrist, and the third placed a looking-glass before his lips, and they found that all traces of respiration and pulsation gradually ceased, insomuch that, after consulting about his condition for some time, they were leaving the room, persuaded that he was really dead, when signs of life appeared, and he slowly revived. He did not die whilst repeating this experiment.

This reviving "by an effort or somehow," seems to be better explained by the hypothesis I have suggested than by any other; namely, that, as in the case of Mr. Holloway, mentioned in the same chapter, his spirit, or soul, was released from his body, but a sufficient rapport maintained to re-unite them.

END OF VOLUME I.

Printed in the United States
By Bookmasters